Programming, Planning & Practice
ARE Mock Exam
(PPP of Architect Registration Exam)

ARE Overview, Exam Prep Tips,
Multiple-Choice Questions and Graphic Vignettes,
Solutions and Explanations

Gang Chen

ArchiteG®, Inc.
Irvine, California

Programming, Planning & Practice ARE Mock Exam (PPP of Architect Registration Exam: ARE Overview, Exam Prep Tips, Multiple-Choice Questions and Graphic Vignettes, Solutions and Explanations

Copyright © 2012 Gang Chen
V1.11   Incorporated minor revisions on 1/1/2015
Cover Photo © 2012 Gang Chen

Copy Editor: Penny L Kortje

ArchiteG®, Inc.
http://www.ArchiteG.com

ISBN: 978-1-61265-006-7

PRINTED IN THE UNITED STATES OF AMERICA

Dedication

To my parents, Zhuixian and Yugen,
my wife, Xiaojie, and my daughters,
Alice, Angela, Amy, and Athena.

Disclaimer

Programming, Planning & Practice ARE Mock Exam (PPP of Architect Registration Exam) provides general information about the Architect Registration Exam. The book is sold with the understanding that neither the publisher nor the authors are providing legal, accounting, or other professional services. If legal, accounting, or other professional services are required, seek the assistance of a competent professional firm.

The purpose of this publication is not to reprint the content of all other available texts on the subject. You are urged to read other materials and tailor them to fit your needs.

Great effort has been taken to make this resource as complete and accurate as possible. However, nobody is perfect and there may be typographical errors or other mistakes present. You should use this book as a general guide and not as the ultimate source on this subject. If you find any potential errors, please send an e-mail to:
info@ArchiteG.com

Programming, Planning & Practice ARE Mock Exam (PPP of Architect Registration Exam) is intended to provide general, entertaining, informative, educational, and enlightening content. Neither the publisher nor the author shall be liable to anyone or any entity for any loss or damages, or alleged loss or damages, caused directly or indirectly by the content of this book.

If you do not wish to be bound by the above, you may return this book to the publisher for a full refund.

Legal Notice

How to Use This Book

We suggest you read *Programming, Planning & Practice ARE Mock Exam (PPP of Architect Registration Exam)* at least three times:

Read once, covering chapter one, two, appendixes, the related FREE PDF files, and other resources. Highlight the information with which you are not familiar.

Read a second time, focusing on memorization of the highlighted information. You can repeat this process as many times as you want until you have mastered the content of the book. Pay special attention to the materials listed in chapter two, section B. **These are the most important documents/publications for the PPP division of the ARE exam.**

After reviewing these materials, take the mock exam, and then check your answers against the answers and explanations in the back. Make sure to read all the explanations for the questions. You may have answered some questions correctly, but find you did so for the wrong reason. Highlight the information you are not familiar with.

Like the real exam, the mock exam includes three types of questions: select the correct answer, check all that apply, and fill in the blank.

Review your highlighted information, and take the mock exam again. Try to answer the questions 100% correctly this time. Repeat the process until you can answer all of the questions correctly.

The mock exam should be taken about two weeks before the real exam, but at least 3 days before the real exam. You should NOT wait until the night before the real exam to practice the mock exam. If you do not do well, you may go into panic mode and NOT have enough time to review your weaknesses.

Read for the final time the night before the real exam. Review ONLY the information you highlighted, especially the questions you did not answer correctly the first time you took the mock exam.

One important tip for passing the graphic vignette section of the ARE PPP division is to become VERY familiar with the commands of the NCARB software. Many people fail the exam simply because they are NOT familiar with this software and cannot finish the graphic vignette section within the exam time limit.

For the graphic vignettes, we include step-by-step solutions, using screen-shots of the NCARB practice program software, so that you can use this book to become familiar with the commands, even when you do NOT have a computer in front of you. This book is also very light to facilitate easy transport. These two features will allow you to review the graphic vignette section whenever you have a few minutes.

All commands are described in an **abbreviated manner**. For example, **Sketch > Line** means go to the menu on the left hand side of your computer screen, click **Sketch,** and then click **Line** to draw a sketch line**.** This is typical for ALL commands throughout the book.

The Table of Contents is very detailed, so you can locate information quickly. If you are on a tight schedule, you can forgo reading the book linearly and jump to the sections you need.

All our books including "ARE Mock Exams Series" and "LEED Exam Guides Series" are available at
GreenExamEducation.com

Check out FREE tips and info at **GeeForum.com**, you can post your questions or vignettes for other users' review and responses.

Table of Contents

Chapter One   Overview of the Architect Registration Exam (ARE)

   1. Important links to FREE and official NCARB documents
   2. A detailed list and brief description of FREE PDF files that can be downloaded from NCARB
      - ARE Guidelines
      - NCARB Education Guidelines
      - Intern Development Program Guidelines
      - IDP Supervisor Guidelines
      - Handbook for Interns and Architects
      - Official exam guide, references index, and practice program (NCARB software) for each ARE division
      - The Burning Question: Why Do We Need ARE Anyway?
      - Defining Your Moral Compass
      - Rules of Conduct

   1. What is IDP?
   2. Who qualifies as an intern?

   1. How to qualify for the ARE
   2. How to qualify for an architect license
   3. What is the purpose of the ARE?
   4. What is NCARB's rolling clock?
   5. How to register for an ARE exam
   6. How early do I need to arrive at the test center?

7. Exam format & time
   - Programming, Planning & Practice
   - Site Planning & Design
   - Building Design & Construction Systems
   - Schematic Design
   - Structural Systems
   - Building Systems
   - Construction Documents and Services
8. How are ARE scores reported?
9. Is there a fixed percentage of candidates who pass the ARE exams?
10. When can I retake a failed ARE division?
11. How much time do I need to prepare for each ARE division?
12. Which ARE division should I take first?
13. ARE exam prep and test-taking tips
14. English system (English or inch-pound units) vs. metric system (SI units)
15. Codes and standards used in this book
16. Where can I find study materials on architectural history?

Chapter Two   **Programming, Planning & Practice (PPP) Division**

   1. A FREE article, *Architectural Programming,* by WBDG (Whole Building Design Guide), a program of the National Institute of Building Science.
   2. *The Architect's Handbook of Professional Practice* (AHPP)
   3. American Institute of Architects (AIA) Documents
   4. *International Building Code* (IBC)
   5. *Architectural Graphic Standards* (AGS)
   6. Access Board, *ADAAG Manual: A Guide to the Americans with Disabilities Act Accessibility Guidelines*
   7. Historic Preservation documents
   8. Construction Specifications Institute (CSI) MasterFormat & *Building Construction*

   1. Overall strategies
   2. Tips
   3. A step-by-step solution to the official NCARB practice program graphic vignette: Site Zoning
   4. Notes on NCARB traps

Chapter One

Overview of the Architect Registration Exam (ARE)

A. First Things First: Go to the website of your architect registration board and read all the requirements for obtaining an architect license in your jurisdiction.
See the following link:
http://www.ncarb.org/Getting-an-Initial-License/Registration-Board-Requirements.aspx

B. Download and Review the Latest ARE Documents at the NCARB Website

1. Important links to FREE and official NCARB documents
The current version of the Architect Registration Exam includes seven divisions:

- Programming, Planning & Practice
- Site Planning & Design
- Building Design & Construction Systems
- Schematic Design
- Structural Systems
- Building Systems
- Construction Documents and Services

Note: Starting July 2010, the 2007 AIA Documents apply to all ARE Exams.

Six ARE divisions have a multiple-choice section and a graphic vignette section. The Schematic Design division has NO multiple-choice section, but two graphic vignette sections.

For the vignette section, you need to complete the following graphic vignette(s) based on the ARE division you are taking:

Programming, Planning & Practice
Site Zoning

Site Planning & Design
Site Grading
Site Design

Building Design & Construction Systems
Accessibility/Ramp
Stair Design
Roof Plan

Schematic Design
Interior Layout
Building Layout

Structural Systems
Structural Layout

Building Systems
Mechanical & Electrical Plan

Construction Documents & Services
Building Section

There is a tremendous amount of valuable information covering every step of becoming an architect available free of charge at the NCARB website:
http://www.ncarb.org/

For example, you can find the education guide regarding professional architectural degree programs accredited by the National Architectural Accrediting Board (NAAB), NCARB's Intern Development Program (IDP) guides, initial license, certification and reciprocity, continuing education, etc. These documents explain how you can qualify to take the Architect Registration Exam.

I find the official ARE Guidelines, exam guide, and practice program for each of the ARE divisions extremely valuable. See the following link:
http://www.ncarb.org/ARE/Preparing-for-the-ARE.aspx

You should definitely start by studying the official exam guide and practice program for the ARE division you are taking.

2. **A detailed list and brief description of the FREE PDF files that can be downloaded from NCARB**
   The following is a detailed list of the FREE PDF files that you can download from NCARB. They are listed in order based on their importance.

   - **ARE Guidelines** includes extremely valuable information on the ARE overview, six steps to complete ARE, multiple-choice section, graphic vignette section, exam format, scheduling, sample exam computer screens, links to other FREE NCARB PDF files, practice software for graphic vignettes, etc. You need to read this at least twice.

   - **NCARB Education Guidelines** (Skimming through this should be adequate)

- **Intern Development Program Guidelines** contains important information on IDP overview, IDP steps, IDP reporting, IDP basics, work settings, training requirements, supplementary education (core), supplementary education (elective), core competences, next steps, and appendices. Most of NCARB's 54-member boards have adopted the IDP as a prerequisite for initial architect licensure. This is why you should be familiar with it. IDP costs $350 for the first three years, and then $75 annually. The fees are subject to change, and you need to check the NCARB website for the latest information. Your IDP experience <u>should be reported to NCARB at least every six months</u> and logged within two months of completing each reporting period (the **Six-Month Rule**). You need to read this document <u>at least twice</u>. It has a lot of valuable information.

- **The IDP Supervisor Guidelines** (Skimming through this should be adequate. You should also forward a copy of this PDF file to your IDP supervisor.)

- **Handbook for Interns and Architects** (Skimming through this should be adequate.)

- **Official exam guide, <u>references index</u>, and practice program (NCARB software) for each ARE division**
  This includes specific information for each ARE division. (Just focus on the documents related to the ARE divisions you are currently taking and read them at least twice. Make sure you install the practice program and become very familiar with it. The real exam is VERY similar to the practice program.)

    a. **Programming, Planning & Practice (PPP)**: Official exam guide and practice program for the PPP division
    b. **Site Planning & Design (SPD)**: Official exam guide and practice program (computer software) for the SPD division
    c. **Building Design & Construction Systems (BDCS)**: Official exam guide and practice program for the BDCS division
    d. **Schematic Design (SD)**: Official exam guide and practice program for the SD division
    e. **Structural Systems (SS)**: Official exam guide, <u>references index</u>, and practice program for the SS division
    f. **Building Systems (BS)**: Official exam guide and practice program for the BS division
    g. **Construction Documents and Services (CDS)**: Official exam guide and practice program for the CDS division

- **The Burning Question: Why Do We Need ARE Anyway?** (Skimming through this should be adequate.)

- **Defining Your Moral Compass** (Skimming through this should be adequate.)

- **Rules of Conduct** is available as a FREE PDF file at:

http://www.ncarb.org/
(Skimming through this should be adequate.)

C. The Intern Development Program (IDP)

1. What is IDP?

IDP is a comprehensive training program jointly developed by the National Council of Architectural Registration Boards (NCARB) and the American Institute of Architects (AIA) to ensure that interns obtain the necessary skills and knowledge to practice architecture <u>independently</u>.

2. Who qualifies as an intern?

Per NCARB, if an individual meets one of the following criteria, s/he qualifies as an intern:
a. Graduates from NAAB-accredited programs
b. Architecture students who acquire acceptable training prior to graduation
c. Other qualified individuals identified by a registration board

D. Overview of the Architect Registration Exam (ARE)

1. How to qualify for the ARE

A candidate needs to qualify for the ARE via one of NCARB's member registration boards, or one of the Canadian provincial architectural associations.

Check with your Board of Architecture for specific requirements.

For example, in California, a candidate must provide verification of a minimum of <u>five</u> years of education and/or architectural work experience to qualify for the ARE.

Candidates can satisfy the five-year requirement in a variety of ways:

- Provide verification of a professional degree in architecture through a program that is accredited by NAAB or CACB.

  OR
- Provide verification of at least five years of educational equivalents.

  OR
- Provide proof of work experience under the direct supervision of a licensed architect

2. **How to qualify for an architect license**

   Again, each jurisdiction has its own requirements. An individual typically needs a combination of about <u>eight</u> years of education and experience, as wells as passing scores on the ARE exams. See the following link:
   http://www.ncarb.org/Reg-Board-Requirements

   For example, the requirements to become a licensed architect in California are:
   - Eight years of post-secondary education and/or work experience as evaluated by the Board (including at least one year of work experience under the direct supervision of an architect licensed in a U.S. jurisdiction or two years of work experience under the direct supervision of an architect registered in a Canadian province)
   - Completion of the Comprehensive Intern Development Program (CIDP) and the Intern Development Program (IDP)
   - Successful completion of the Architect Registration Examination (ARE)
   - Successful completion of the California Supplemental Examination (CSE)

   California does NOT require an accredited degree in architecture for examination and licensure. However, many other states do.

3. **What is the purpose of the ARE?**

   The purpose of ARE is NOT to test a candidate's competency on every aspect of architectural practice. Its purpose is to test a candidate's competency on providing professional services to protect the <u>health, safety, and welfare</u> of the public. It tests candidates on the <u>fundamental</u> knowledge of pre-design, site design, building design, building systems, and construction documents and services.

   The ARE tests a candidate's competency as a "specialist" on architectural subjects. It also tests her abilities as a "generalist" to coordinate other consultants' works.

   You can download the exam content and references for each of the ARE divisions at the following link:
   http://www.ncarb.org/are/40/StudyAids.html

4. **What is NCARB's rolling clock?**
   a. Starting on January 1, 2006, a candidate MUST pass ALL ARE sections within five years. A passing score for an ARE division is only valid for five years, and a candidate has to retake this division if she has NOT passed all divisions within the five year period.

   b. Starting on January 1, 2011, a candidate who is authorized to take ARE exams MUST take at least one division of the ARE exams within five years of the authorization. Otherwise, the candidate MUST apply for the authorization to take ARE exams from an NCARB member board again.

   These rules were created by the **NCARB's rolling clock** resolution and passed by NCARB council during the 2004 NCARB Annual Meeting.

5. **How to register for an ARE exam**
   Go to the following website and register:
   http://www.prometric.com/NCARB/default.htm

6. **How early do I need to arrive at the test center?**
   Be at the test center at least 30 minutes BEFORE your scheduled test time, OR you may lose your exam fee.

7. **Exam format & time**
   All ARE divisions are administered and graded by computer. Their detailed exam format and time allowances are as follows:

   1) Programming, Planning & Practice (PPP)

Introduction Time:	15 minutes	
MC Testing Time:	**2 hours**	**85 items**
Scheduled Break:	15 minutes	
Introduction Time:	15 minutes	
Graphic Testing Time:	**1 hour**	**Site Zoning (1 vignette)**
Exit Questionnaire:	15 minutes	
Total Time	**4 hours**	

   2) Site Planning & Design (SPD)

Introduction Time:	15 minutes	
MC Testing Time:	**1.5 hours**	**65 items**
Scheduled Break:	15 minutes	
Introduction Time:	15 minutes	
2 Graphic Vignettes:	**2 hours**	**Site Grading, Site Design**
Exit Questionnaire:	15 minutes	
Total Time	**4.5 hours**	

   3) Building Design & Construction Systems (BDCS)

Introduction Time:	15 minutes	
MC Testing Time:	**1.75 hours**	**85 items**
Scheduled Break:	15 minutes	
Introduction Time:	15 minutes	
3 Graphic Vignettes:	**2.75 hours**	**Accessibility/Ramp, Stair Design, Roof Plan**
Exit Questionnaire:	15 minutes	
Total Time	**5.5 hours**	

4) Schematic Design (SD)

Introduction Time:	15 minutes	
Graphic Testing Time:	**1 hour**	**Interior Layout (1 vignette)**
Scheduled Break:	15 minutes	
Introduction Time:	15 minutes	
Graphic Testing Time:	**4 hours**	**Building Layout (1 vignette)**
Exit Questionnaire:	15 minutes	
Total Time	**6 hours**	

5) Structural Systems (SS)

Introduction Time:	15 minutes	
MC Testing Time:	**3.5 hours**	**125 items**
Scheduled Break:	15 minutes	
Introduction Time:	15 minutes	
Graphic Testing Time:	**1 hour**	**Structural Layout (1 vignette)**
Exit Questionnaire:	15 minutes	
Total Time	**5.5 hours**	

6) Building Systems (BS)

Introduction Time:	15 minutes	
MC Testing Time:	**2 hours**	**95 items**
Scheduled Break:	15 minutes	
Introduction Time:	15 minutes	
Graphic Testing Time:	**1 hour**	**Mechanical & Electrical Plan (1 vignette)**
Exit Questionnaire:	15 minutes	
Total Time	**4 hours**	

7) Construction Documents and Services (CDS)

Introduction Time:	15 minutes	
MC Testing Time:	**2 hours**	**100 items**
Scheduled Break:	15 minutes	
Introduction Time:	15 minutes	
Graphic Testing Time:	**1 hour**	**Building Section (1 vignette)**
Exit Questionnaire:	15 minutes	
Total Time	**4 hours**	

8. How are ARE scores reported?

All ARE scores are reported as Pass or Fail. ARE scores are processed within 4 to 6 weeks, and sent to your Board of Architecture. Your board then does additional processing and forwards the scores to you.

9. Is there a fixed percentage of candidates who pass the ARE exams?

No, there is NOT a fixed percentage of passing or failing. If you meet the minimum competency required to practice as an architect, you pass. The passing scores are the same for all Boards of Architecture.

10. When can I retake a failed ARE division?

You can only take the same ARE division once within a 6-month period.

11. How much time do I need to prepare for each ARE division?

Every person is different, but on average you need about 40 hours to prepare for each ARE division. You need to set a realistic study schedule and stick with it. Make sure you allow time for personal and recreational commitments. If you are working full time, my suggestion is that you allow no less than 2 weeks but NOT more than 2 months to prepare for each ARE division. You should NOT drag out the exam prep process too long and risk losing your momentum.

12. Which ARE division should I take first?

This is a matter of personal preference, and you should make the final decision.

Some people like to start with the easier divisions and pass them first. This way, they build more confidence as they study and pass each division.

Other people like to start with the more difficult divisions so that if they fail, they can keep busy studying and taking the other divisions while the clock is ticking. Before they know it, six months has passed and they can reschedule if need be.

Programming, Planning & Practice (PPP) and Building Design & Construction Systems (BDCS) divisions often include some content from the Construction Documents and Service (CDS) division. It may be a good idea to start with CDS and then schedule the exams for PPP and BDCS soon after.

13. ARE exam prep and test-taking tips

You can start with Construction Documents and Services (CDS) and Structural Systems (SS) first because both divisions give a limited scope, and you may want to study building regulations and architectural history (especially famous architects and buildings that set the trends at critical turning points) before you take other divisions.

Complete mock exams and practice questions and vignettes, including those provided by NCARB's practice program and this book, to hone your skills.

Form study groups and learn the exam experience of other ARE candidates. The forum at our website is a helpful resource. See the following link:
http://GreenExamEducation.com/

Take the ARE exams as soon as you become eligible, since you probably still remember portions of what you learned in architectural school, especially structural and architectural history. Do not make excuses for yourself and put off the exams.

The following test-taking tips may help you:
- Pace yourself properly. You should spend about one minute for each Multiple-Choice (MC) question, except for the SS division questions which you can spend about one and a half minutes on.
- Read the questions carefully and pay attention to words like *best, could, not, always, never, seldom, may, false, except,* etc.
- For questions that you are not sure of, eliminate the obvious wrong answer and then make an educated guess. Please note that if you do NOT answer the question, you automatically lose the point. If you guess, you at least have a chance to get it right.
- If you have no idea what the correct answer is and cannot eliminate any obvious wrong answers, then do not waste too much time on the question and just guess. Try to use the same guess answer for all of the questions you have no idea about. For example, if you choose "d" as the guess answer, then you should be consistent and use "d" whenever you have no clue. This way, you are likely have a better chance at guessing more answers correctly.
- Mark the difficult questions, answer them, and come back to review them AFTER you finish all MC questions. If you are still not sure, go with your first choice. Your first choice is often the best choice.
- You really need to spend time practicing to become VERY familiar with NCARB's graphic software and know every command well. This is because the ARE graphic vignette is a timed test, and you do NOT have time to think about how to use the software during the test. If you do not know how, you will NOT be able to finish your solution to the vignette on time.
- The ARE exams test a candidate's competency to provide professional services protecting the <u>health, safety, and welfare</u> of the public. Do NOT waste time on aesthetic or other design elements not required by the program.

ARE exams are difficult, but if you study hard and prepare well, combined with your experience, IDP training, and/or college education, you should be able to pass all divisions and eventually be able to call yourself an architect.

14. English system (English or inch-pound units) vs. metric system (SI units)
This book is based on the English system or English units; the equivalent value in metric system or SI units follows in parentheses. All SI dimensions are in millimeters unless noted otherwise. The English or inch-pound units are based on the module used in the U.S. The SI units noted are simple conversions from the English units for information only and are not necessarily according to a metric module.

15. Codes and standards used in this book

We use the following codes and standards:

American Institute of Architects, Contract Documents, Washington, DC.

Canadian Construction Documents Committee, CCDC Standard Documents, 2006, Ottawa.

16. Where can I find study materials on architectural history?

Every ARE exam may have a few questions related to architectural history. The following are some helpful links to FREE study materials on the topic:

http://www.essentialhumanities.net/arch.php
http://issuu.com/motimar/docs/history_synopsis?viewMode=magazine
http://www.ironwarrior.org/ARE/Materials_Methods/m_m_notes_2.pdf

Chapter Two

Programming, Planning & Practice (PPP) Division

A. General Information

1. Exam content

An architect should have the skills and knowledge of project development; architectural programming; building codes and regulations; economic, social and environmental issues; and practice and project management.

The exam content for the PPP division of the ARE includes:

1) **Programming & Analysis (24% to 30%)**
   - Architectural Programming
   - Interpreting Existing Site/Environmental Conditions and Data
   - Adaptive Reuse of Buildings and/or Materials
   - Space Planning and Facility Planning/Management
   - Fixtures, Furniture, Equipment, and Finishes

2) **Environmental Social & Economic Issues (23% to 29%)**
   - Regional Impact on Project
   - Community-Based Awareness
   - Hazardous Conditions and Materials
   - Design Principles
   - Alternative Energy Systems, New Technologies, and Sustainable Design
   - Architectural History and Theory

3) **Codes & Regulations (10% to 13%)**
   - Government and Regulatory Requirements and Permit Processes
   - Adaptive Reuse of Buildings and/or Materials
   - Specialty Codes and Regulations including Accessibility Laws, Codes and Guidelines

4) **Project & Practice Management (33% to 39%)**
   - Project Delivery Methods
   - Project Budget Management
   - Project Schedule Management
   - Contracts for Professional Services and Contract Negotiation
   - Construction Procurement Processes
   - Risk Management and Legal Issues Pertaining to Practice and Contracts

For the graphic vignette, you will be required to define areas suitable for building and other site improvement per the program. You need to delineate a maximum buildable envelope and site profile according to environmental constraints and zoning regulations.

2. Official exam guide and practice program for the PPP division

You need to read the official exam guide for the PPP division at least twice. Make sure you install the PPP division practice program on your computer and become very familiar with it. The real exam is VERY similar to the practice program.

You can download the official exam guide and practice program for the PPP division at the following link:
http://www.ncarb.org/en/ARE/Preparing-for-the-ARE.aspx

B. Important Documents and Publications for the PPP Division of the ARE Exam

PPP is one of the ARE divisions which is very hard to prepare for by simply reading a finite number of books within a short amount of time. This is because many of the questions in this division are based on work experience, and may also include questions from other ARE divisions.

We shall help you alleviate this problem by bringing your attention to some of the most common issues in architectural practice that relate to the PPP division.

The PPP division may include questions from a broad scope, but the questions **have to** be issues an **average** architect would encounter during normal architectural practice. NCARB may include some obscure issues, but not too many. Otherwise, the ARE tests would NOT be **legally defensible**.

Based on our research, the most important documents/publications for the PPP division of the ARE exam are:

1. A FREE article, *Architectural Programming*, by WBDG (Whole Building Design Guide), a program of National Institute of Building Science.

See the following link:
http://www.wbdg.org/design/dd_archprogramming.php

2. *The Architect's Handbook of Professional Practice* (AHPP)
Demkin, Joseph A., AIA, Executive Editor. *The Architect's Handbook of Professional Practice* (AHPP). The American Institute of Architects & Wiley, latest edition.

This comprehensive book covers all aspects of architectural practice, and includes two CDs containing sample AIA contract documents. You may have about ten real ARE PPP division questions based on this publication. Therefore look through this book a few times and know some of the basic architectural practice elements. Read the related portions carefully, such as delivery methods and compensation (design-bid-build, construction management, and design-build), and contracts and agreements.

3. **American Institute of Architects (AIA) Documents**
   You may have fifteen to twenty real ARE PPP division questions based on AIA documents. The questions will deal with the front end stuff of a project, i.e., issues related to programming, planning & practice like scope of work and contract, and NOT the backend stuff, i.e., constructions administration issues like job site observation, submittals, and shop drawings.

   Reading the summary of the AIA Documents is NOT adequate preparation. You need to read the actual text. Fortunately, you do NOT have to read all the available AIA documents.

   Three possible study solutions are as follows:
   a. Buy ONLY the AIA documents you need from your local AIA office. The AIA documents listed below are important, especially those in **bold** font, read them at least three times. You may have **many** real ARE PPP division questions based on the following AIA documents listed in **bold** font:

   * **A101–2007, Standard Form of Agreement Between Owner and Contractor where the basis of payment is a Stipulated Sum (CCDC Document 2)**
   * **A201–2007, General Conditions of the Contract for Construction**
   * A503–2007, Guide for Supplementary Conditions
   * A701–1997, Instructions to Bidders (CCDC Document 23)
   * **B101–2007 (Former B141–1997), Standard Form of Agreement Between Owner and Architect (RAIC Document 6)**

     You can find FREE sample forms and commentaries for AIA documents A201 & B101 at the following link:
     http://www.aia.org/contractdocs/aiab081438

   * **C401–2007 (Former C141–1997), Standard Form of Agreement Between Architect and Consultant**

     There are some <u>major changes</u> between C401-2007 and the older version C141–1997.

     However, a FREE version of sample C141–1997 is available at:
     https://app.ncarb.org/are/StudyAids/_C141.pdf

   * G701–2001, Change Order
   * G702–1992, Application and Certificate for Payment
   * G704–2000, Certificate of Substantial Completion

   AIA updates their documents roughly every 10 years. Although, please note that AIA does NOT update <u>all</u> available documents at the same time. For example, A701–1997, Instructions to Bidders (CCDC Document 23) is still the most current form.

See the AIA documents price list at the following link for the latest edition of AIA documents:
http://www.aia.org/aiaucmp/groups/aia/documents/pdf/aias076346.pdf

  b. Buy AHPP which has a CD including the sample AIA documents. AHPP itself is also a very important publication for the PPP division.

  c. After you register with NCARB for the ARE exams and log in to their site, you have FREE access to the AIA documents online.

4. *International Building Code* **(IBC)**
   International Code Council, Inc. (ICC, 2006), *International Building Code* (IBC).
   You may have about ten real ARE PPP division questions based on this publication. You need to become familiar with some of the more commonly used code sections, such as: allowable areas and allowable areas increase, unlimited areas, egresses, width and numbers of exits required, minimum exit passage width, occupancy groups and related exit occupancy load factors, types of construction, minimum number of required plumbing fixtures required, etc.

   See the following link for some FREE IBC code sections citations:
   http://publicecodes.citation.com/icod/ibc/2006f2/index.htm?bu=IC-P-2006-000001&bu2=IC-P-2006-000019

5. *Architectural Graphic Standards* **(AGS)**
   Ramsey, Charles George, and John Ray Hoke Jr. *Architectural Graphic Standards.*
   The American Institute of Architects & Wiley, latest edition.

   There may be a few questions asking you to identify some of the basic graphic symbols. This is a good book to skim through.

6. **Access Board, *ADAAG Manual: A Guide to the Americans with Disabilities Accessibility Guidelines*. East Providence, RI: BNI Building News. ADA Standards for Accessible Design are available as FREE PDF files at:**

   http://www.ada.gov/

   AND
   http://www.access-board.gov/adaag/html/figures/index.html

7. **Historic Preservation documents**
   Based on examinee feedback, there are a few questions regarding historic preservation in the PPP division exam.

   You should **read** the following documents **at least twice** to become familiar with them. Pay special attention to information in italics and the shaded areas of the PDF file:

- The two FREE PDF files are *The Secretary of the Interior's Standards for the Treatment of Historic Properties with Guidelines for Preserving, Rehabilitating Restoring & Reconstructing Historic Buildings* and *The Secretary of the Interior's Standards for Rehabilitation & Illustrated Guidelines for Rehabilitating Historic Buildings- Standards* available at:
http://www.ironwarrior.org/ARE/Historic_Preservation/

AND
http://www.nps.gov/hps/tps/tax/rhb/

You should **look through** the following document and become familiar with it:
- ***National Historic Preservation Act (NHPA)*** at the following link:
http://www.gsa.gov/portal/content/104441

8. **Construction Specifications Institute (CSI) MasterFormat & *Building Construction***
Become familiar with the new 6-digit CSI Construction Specifications Institute (CSI) MasterFormat as there may be a few questions based on this publication. Make sure you know which items/materials belong to which CSI MasterFormat specification section, and memorize the major section names and related numbers. For example, Division 9 is Finishes, and Division 5 is Metal, etc. Another one of my books, *Building Construction*, has detailed discussions on CSI MasterFormat specification sections.

Mnemonics for the 2004 CSI MasterFormat

The following is a good mnemonic, which relates to the 2004 CSI MasterFormat division names. Bold font signals the gaps in the numbering sequence.

This tool can save you lots of time: if you can remember the four sentences below, you can easily memorize the order of the 2004 CSI MasterFormat divisions. The number sequencing is a bit more difficult, but can be mastered if you remember the five bold words and numbers that are not sequential. Memorizing this material will not only help you in several divisions of the ARE, but also in real architectural practice

Mnemonics (pay attention to the underlined letters):
Good students can memorize material when teachers order.
F students earn F's simply 'cause **forgetting** principles have **an** effect. (21 and 25)
C students **end** everyday understanding things without memorizing. (31)
Please make professional pollution prevention inventions **everyday**. (40 and 48)

1-Good.................................. General Requirements
2-Students............................. (Site) now Existing Conditions
3-Can.....................................Concrete
4-Memorize...........................Masonry
5-Material ....Metals
6-When..................................Woods and Plastics
7-Teachers.............................Thermal and Moisture
8-Order..................................Openings

9-F...Finishes
10-Students............................Specialties
11-Earn..................................Equipment
12-F's.....................................Furnishings
13-Simply..............................Special Construction
14-'Cause...............................Conveying
21-Forgetting .... **Fire**
22-Principles.........................Plumbing
23-Have.................................HVAC
25-An.................................... **Automation**
26-Effect................................Electric

27-C......................................Communication
28-Students............................Safety & Security
31-End.................................. **Earthwork**
32-Everyday...........................Exterior
33-Understanding ....Utilities
34-Things...............................Transportation
35-Without Memorizing........Waterways and Marine

40-Please.............................. **Process Integration**
41-Make.................................Material Processing and Handling Equipment
42-Professional.....................Process Heating, Cooling, and Drying Equipment
43-Pollution...........................Process Gas and Liquid Handling, Purification and Storage
 Equipment
44-Prevention........................Pollution Control Equipment
45-Inventions........................Industry-Specific Manufacturing Equipment
48-Everyday..........................**Electrical Power Generation**
Note:
There are 49 CSI divisions. The "missing" divisions are those "reserved for future expansion" by CSI. They are intentionally omitted from the list.

C. Overall Strategies and Tips for NCARB Graphic Vignettes

1. Overall strategies

To most candidates, the Multiple Choice (MC) portion of an ARE division is harder than the graphic vignette portion. Some of the MC questions are based on experience and you do NOT have a set of fixed study materials for them. You WILL make some mistakes on the MC questions no matter how hard you study.

On the other hand, the graphic vignettes are relatively easier, and there are good ways to prepare for them. You should really take the time to study and practice the NCARB graphic software well, and try to **nail all the graphic vignettes** perfectly. This way, you will have a better chance to pass even if you answer some MC questions incorrectly.

Tips: Most people do poorly on the MC portion of the PPP division, especially those who do NOT have a lot of working experience, but curiously not too many people fail because of the MC portion. Most people fail the ENTIRE PPP section because they have made **one** *fatal mistake on the graphic vignette section. So, practice the NCARB PPP practice program graphic software and make sure you absolute NAIL the vignette section. This is a key for you to pass.*

The official NCARB PPP exam guide gives a passing and failing solution the sample vignette, but it does NOT show you the step-by-step details.

We are going to fill in the blanks here and offer you step-by-step instructions, command by command.

You really need to spend time practicing to become VERY familiar with NCARB's graphic software. This is because all ARE graphic vignettes are timed, and you do NOT have the luxury to think about how to use the software during the exam. Otherwise, you may NOT be able to finish your solution on time.

The following solution is based on the official NCARB PPP practice program for the **ARE 4.0**. Future versions of ARE may have some minor changes, but the principles and fundamental elements should be the same. The official NCARB PPP practice program has not changed much since its introduction and the earlier versions are VERY similar to, if not exactly the same as, the current ARE 4.0. The actual graphic vignette of the PPP division should be VERY, VERY similar to the practice one on NCARB's website.

2. Tips

1) You need to install the NCARB PPP practice program, and become familiar with it. I am NOT going to repeat the vignette description and requirements here since they are already written in the NCARB practice program.

See the following link for a FREE download of the NCARB practice program: http://www.ncarb.org/ARE/Preparing-for-the-ARE.aspx

2) Review the general test directions, vignette directions, program, and tips carefully.

3) Press the space bar to go to the work screen.

4) Read the program and codes in the NCARB Exam Guide several times the week before your exam. Become VERY familiar with this material, and you will be able to read the problem requirements MUCH faster during the real exam because you can immediately identify which criteria are different from the practice exam.

3. A step-by-step solution to the official NCARB practice program graphic vignette: Site Zoning

1) Surface improvements are prohibited within 5 ft of any property line. Use **Sketch > Rectangle** to draw a number of rectangles defining the no improvement areas within 5 ft of all property lines (figure 2.1).

2) Construction of buildings and other surface improvements is prohibited within 25 ft of the lake high water line. Use **Sketch > Circle** to draw a number of circles with 25 ft radii defining the no improvement areas within 25 ft of the lake high water line. The centers of the circles should be placed on the lake high water line (figure 2.2).

Note: The radius of the circle displays on the lower-left-hand corner of the screen when drawn. After you draw the first circle, the radius for rest of the circles will stay at 25 ft, and you just need to click on the lake high water line to place them.

3) Use **Draw > Secondary Construction Area** to draw the surface improvement areas (figure 2.3).

Note: The secondary construction area is a polygon. You draw a starting point, click for each corner of the polygon, and then close the polygon by clicking on the starting point again.

4) Use **Sketch > Rectangle** to draw a number of rectangles defining the building setbacks from all property lines (figure 2.4). Front yard setbacks are only considered from Main Street.
Front yard setbacks from property line along Main Street: 25 ft
Rear yard setbacks: 30 ft
Side yard setbacks: 10 ft

5) Construction of buildings is prohibited within the existing drainage easement. Use **Draw > Buildable Area** to draw the buildable areas (figure 2.5).

Note: The building construction area is also a polygon and drawn as described in step three.

In this case, the existing drainage easement is MORE restrictive than the 10 ft side yard setback requirement, so we comply with the more restrictive requirement.

6) Draw the profile of the existing grade at Section A-A:
   - Use **Sketch** > **Line** to project the intersections of the contour lines and section line A-A to the view below (figure 2.6).
   - Per the elevations of the contour lines in plan, use **Draw** > **Grade** to draw the profile of the existing grade at Section A-A on the grid (figure 2.7).

Note: The key of solving this vignette is locating the benchmark and its elevation. The benchmark is located at the intersection of the lake high water line and the west property line of Lot A. Its elevation is 105', the same as the lake high water line.

7) The maximum building height limit within 65 ft of the west property line of Lot A shall be 45 ft above the benchmark elevation. Since the benchmark elevation is 105', we can calculate this as: 105' + 45' = 150'. Use **Sketch** > **Rectangle** to draw a rectangle showing this criterion (figure 2.8).

8) The maximum building height limit between 0 ft and 40 ft of the east property line in Lot B shall be 20 ft above the grade at the property line. The grade at the intersection of the east property line in Lot B and the section line of Section A-A is 145'. We can calculate this as: 145' + 20' = 165'. Use **Sketch** > **Rectangle** to draw a rectangle showing this requirement (figure 2.9).

9) The maximum building height limit shall be 80 ft above the benchmark elevation. Since the benchmark elevation is 105', we can calculate this as: 105' + 80' = 185'. Use **Sketch** > **Rectangle** to draw a rectangle showing this criterion (figure 2.10).

10) The maximum building envelope is restricted to an elevation defined by a 30-degree line rising eastward from a point at an elevation of 20 ft directly above the benchmark. Since the benchmark elevation is 105', we can calculate this as: 105' + 20' = 125'. Use **Sketch** > **Line** to draw a line showing this criterion (figure 2.11).

Note: The angle of the line displays on the lower-left-hand corner as you draw. Getting the angle exactly at 30 degree is almost impossible, because of NCARB's software limitation. The actual angle we draw is 29.79 degrees, which is acceptable and within the allowable range of discrepancy. The angle should be drawn smaller than 30 degrees, rather than larger, because it is more restrictive and on the safe side.

11) Use **Sketch** > **Line** to draw lines projecting down from the buildable areas *at section line A-A* in the floor plan. (figure 2.12).

Note: The program explicitly states, "On the grid, draw the profile of the maximum building envelope for each lot at Section A-A," NOT at the side yard setback lines.

It is VERY easy to make the mistake of drawing the building profile all the way to the side yard setback lines, especially after NCARB requires you to draw the 30-degree line from the property line.

12) Use **Draw** > **Building Profile** to draw the building profiles which comply with all the criteria listed in steps 7 to 11 (figure 2.13).

13) Use **Sketch > Hide Sketch Elements** to hide sketch elements. This is the final solution (figure 2.14).

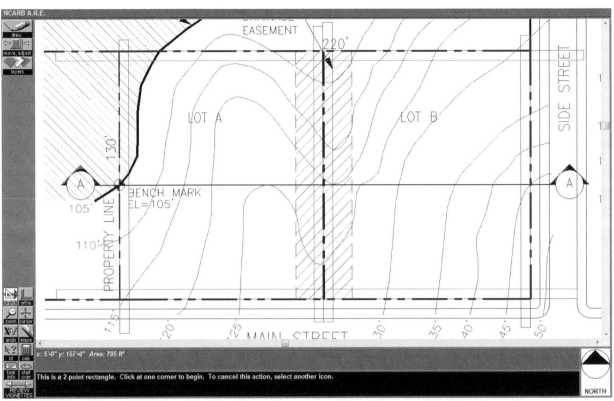

Figure 2.1   Use **Sketch > Rectangle** to draw a number of rectangles defining the no improvement areas within 5 ft of all property lines.

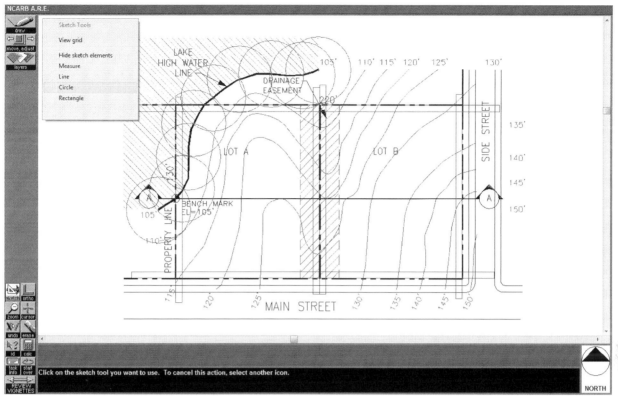

Figure 2.2   Use **Sketch > Circle** to draw a number of circles with 25 ft radii defining the no improvement areas within 25 ft of the lake high water line.

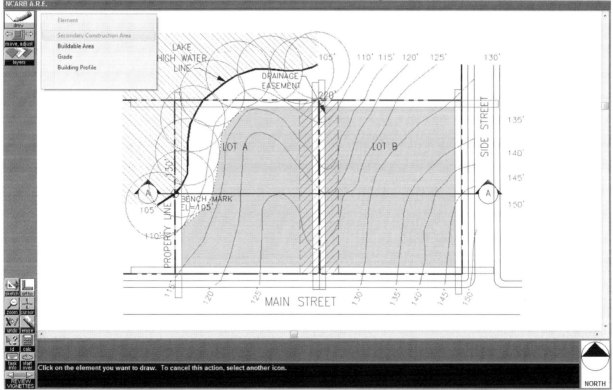

Figure 2.3   Use **Draw > Secondary Construction Area** to draw the surface improvement areas.

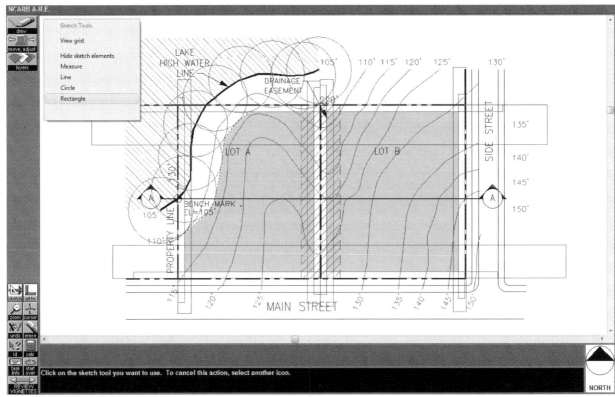

Figure 2.4   Use **Sketch > Rectangle** to draw a number of rectangles defining the building setbacks from all property lines.

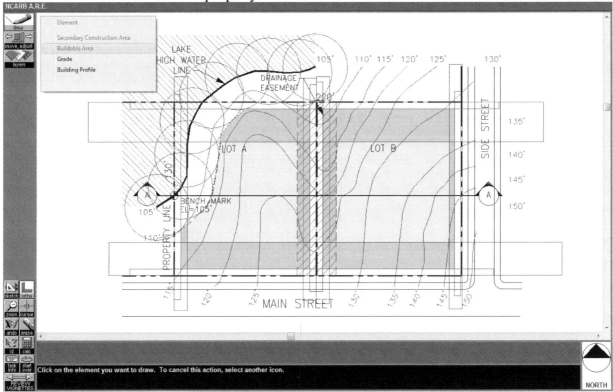

Figure 2.5   Use **Draw > Buildable Area** to draw the buildable areas.

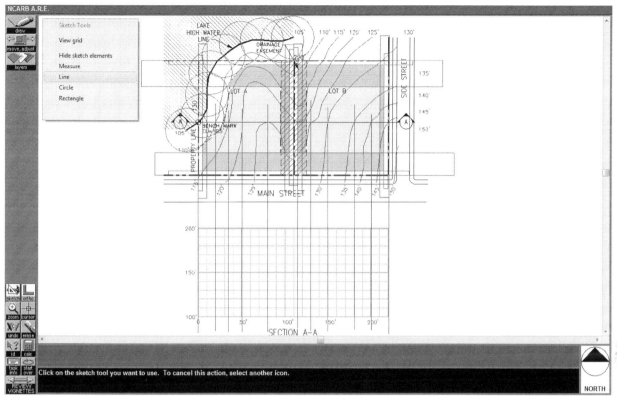

Figure 2.6   Use **Sketch > Line** to project the intersections of the contour lines and section line A-A to the grid below.

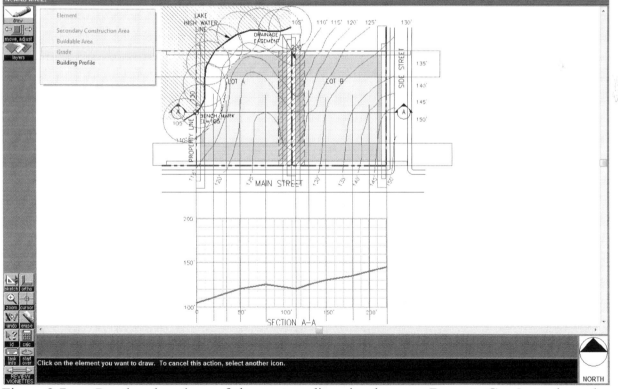

Figure 2.7   Per the elevations of the contour lines in plan, use **Draw > Grade** to draw the profile of the existing grade at Section A-A on the grid.

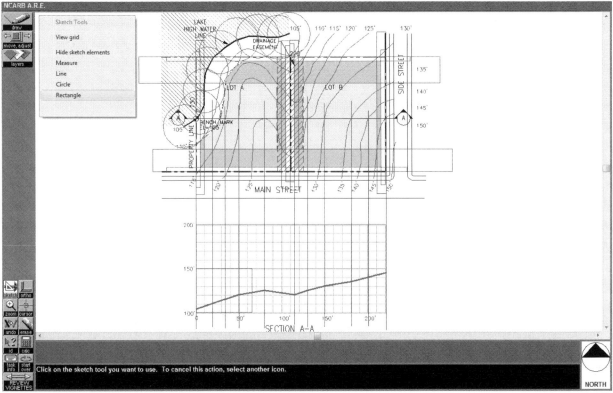

Figure 2.8   Use **Sketch > Rectangle** to draw a rectangle showing the maximum building height limit within 65 ft of the west property line of Lot A.

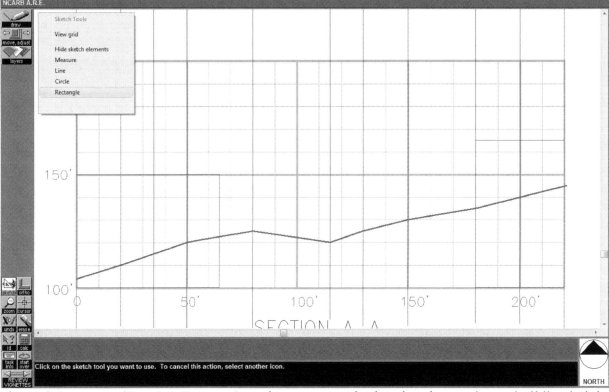

Figure 2.9   Use **Sketch > Rectangle** to draw a rectangle showing the maximum building height limit between 0 ft and 40 ft of the east property line in Lot B.

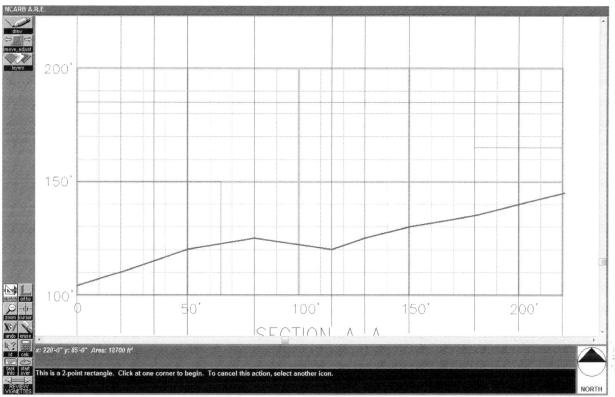

Figure 2.10   Use **Sketch > Rectangle** to draw a rectangle showing that the maximum building height limit shall be 80 ft above the benchmark elevation.

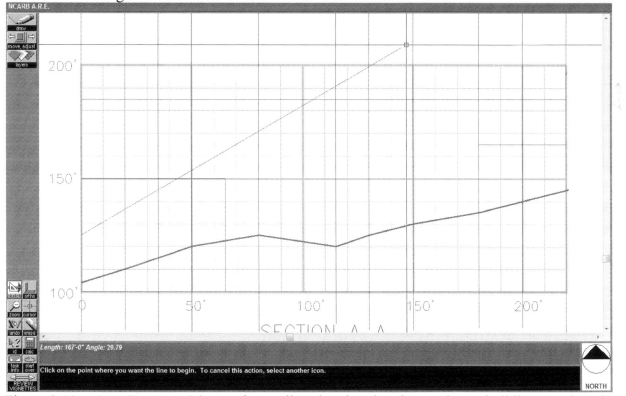

Figure 2.11   Use **Sketch > Line** to draw a line showing that the maximum building envelope is restricted to an elevation defined by a 30-degree line.

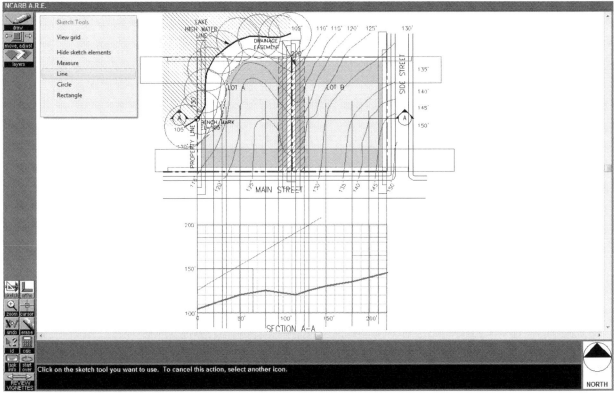

Figure 2.12   Use **Sketch > Line** to draw lines projecting down from the buildable areas at section line A-A in the floor plan.

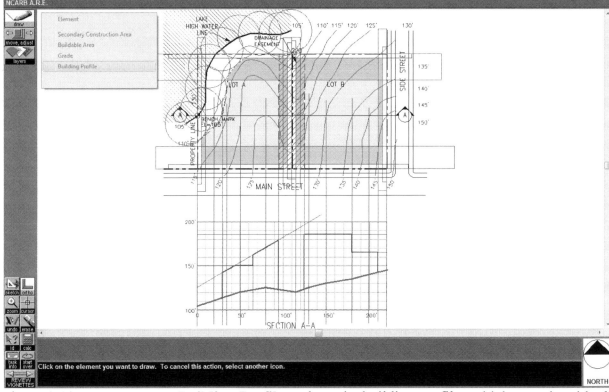

Figure 2.13   Use **Draw > Building Profile** to draw the building profiles which comply with all the criteria listed in steps 7 to 11.

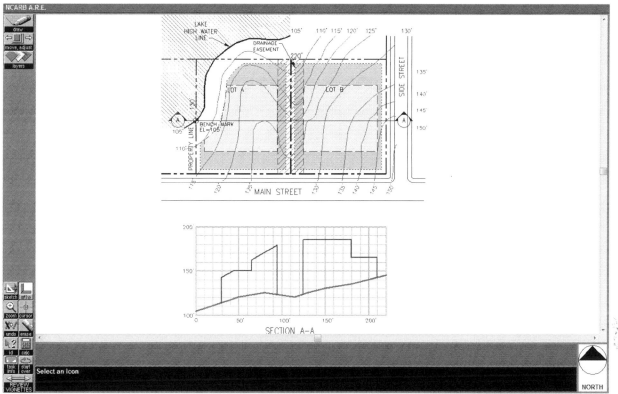

Figure 2.14   Use **Sketch > Hide Sketch Elements** to hide sketch elements. This is the final solution.

4. Notes on NCARB traps

Several **common errors** or **traps** which NCARB wants you to fall into are:

1) The **existing drainage easement** is MORE restrictive than the 10-ft side yard setback requirement, so you need to comply with the more restrictive requirement for the buildable areas and building profiles.

2) The existing drainage easement area can still be a **secondary construction area**.

3) The key of solving this vignette is locating the **benchmark** and its elevation. The benchmark is located at the intersection of the lake high water line and the west property line of Lot A. Its elevation is 105', the same as the lake high water line.

4) The program states, "On the grid, draw the profile of the maximum building envelope for each lot **at Section A-A**," NOT at the side yard setback lines. It is VERY easy to make the mistake of drawing the building profile all the way to the side yard setback lines, especially after NCARB requires you to draw the 30-degree height restriction angle from a property line.

Chapter Three

ARE Mock Exam for
Programming, Planning & Practice (PPP) Division

A. Mock Exam: PPP Multiple-Choice (MC) Section

1. Which of the following is true about lead-based paints in existing U.S. buildings? **Check the two that apply.**
   a. Lead-based paints have been banned in all U.S. buildings.
   b. Lead-based paints have been banned in all new U.S. residential construction.
   c. Lead-based paints pose a greater health risk to children seven years or younger.
   d. Lead-based paints pose a greater health risk to the elderly.

2. Which of the following consultants is most likely to be involved with the preliminary design of a project? **Check the two that apply.**
   a. mechanical engineer
   b. electrical engineer
   c. plumbing engineer
   d. landscape architect
   e. civil engineer

3. Which of the following generates the most indoor air quality problems?
   a. lack of air conditioning equipment
   b. poor construction practice
   c. soils brought in by pedestrian traffic
   d. poor ventilation
   e. no natural ventilation

4. Two construction firms are working on the same building with one owner. One firm is working with the owner on the building shell, and the other is working with the owner on the interior improvement under a separate contract. This arrangement is known as
   a. a multiple prime
   b. an associated firm
   c. a joint venture
   d. partnering

5. A homeowner asks an architect's opinion on whether to renovate his home or to completely demolish it and build a new one. The architect should suggest the homeowner to:
    a. compare the cost of the two choices
    b. seek LEED certification for the home
    c. complete a feasibility study
    d. complete a life safety study

6. EPA identified maximum sound level to protect against hearing loss and other disruptive effects from noise, such as sleep disturbance, stress, and learning detriment as
    a. 20 db
    b. 50 db
    c. 70 db
    d. 80 db

7. Blocking and stacking are
    a. terms used in masonry construction
    b. terms used in programming
    c. terms used in structural calculations
    d. terms used in design development

8. Which of the following is the most common method of estimating construction cost at schematic design stage?
    a. unit-area cost
    b. time and materials
    c. fixed fee
    d. value engineering

9. What does the first C in CC&R stand for?
    a. Conditions
    b. Covenants
    c. Cost
    d. Codes

10. Square-foot [square-meter] costs of similar buildings is a common method of building construction cost estimates at
    a. the programming stage
    b. the design development stage
    c. the construction document stage
    d. bidding

11. An operating pro-forma is (Check the two that apply.)
    a. a year-by-year projection of a project's expenses and income
    b. a detailed breakdown of a project's operation expense
    c. a return on investment ratio
    d. of particular interest to a project's lenders and investors

12. Which of the following can cause mold inside a building wall? **Check the three that apply.**
    a. poor ventilation
    b. poor drainage
    c. flashing
    d. organic feedstock
    e. EIFS

13. An architectural project program should include which of the following? **Check the four that apply.**
    a. a basis of design
    b. owner project requirements
    c. type of structural system
    d. type of HVAC system
    e. a budget
    f. type and quantity of spaces

14. The project schedule in figure 3.1 on the following page is known as a
    a. Gantt chart
    b. critical path method (CPM)
    c. program evaluation and review technique (PERT)
    d. project cycle method (PCM)

15. What is the total numbers of days needed to finish the project per the project schedule in figure 3.1?
    a. 24
    b. 32
    c. 36
    d. 38

16. The dashed arrows in figure 3.1 are known as
    a. knots
    b. paths
    c. dummies
    d. processes

17. Which of the following is not a part of architectural programming? **Check the two that apply.**
    a. functional and operational requirements
    b. scoping
    c. decision-making processes
    d. codes and regulations
    e. geotechnical surveys

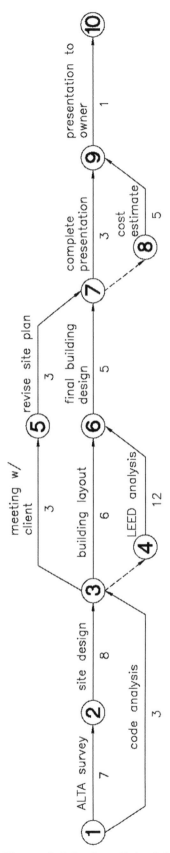

Figure 3.1 Project Schedule

18. The best time to make changes to a project is during the
    a. programming stage
    b. schematic design stage
    c. design development stage
    d. construction documents stage
    e. construction administration stage

19. An architect is working on a major building in a shopping center project. She is setting up a project programming meeting. Who should be invited to the meeting? **Check the four that apply.**
    a. the owners
    b. the building officials
    c. the tenants
    d. the design consultants
    e. the leasing team
    f. the neighbors

20. Which of the following site components can have an unobvious high cost impact at a given site?
    a. topography
    b. utility availability
    c. traffic
    d. vegetation and wildlife
    e. zoning requirements

21. Building efficiency is _____.

22. Tare space is _____.

23. Which of the following are essential elements for architectural programming? **Check the three that apply.**
    a. topography
    b. project type
    c. goals and objectives
    d. quantitative requirements
    e. vegetation and wildlife
    f. zoning requirements

24. After the award of a construction contract, which of the following shall the contractor submit for architect review?
    a. a punch list
    b. a schedule of construction
    c. a list of potential change orders
    d. a change order log
    e. a submittal log

25. According to the Americans with Disabilities Act (ADA) [CSA-B651-04] Accessibility Guidelines, the minimum clear passage width for single wheelchair may be reduced to
    a. 32"
    b. 36"
    c. 44"
    d. 48"

26. Which of the following types of estimates is the most accurate?
    a. unit price
    b. an estimate based on the construction cost of similar buildings
    c. cost of building systems
    d. historical data of the same type of construction

27. An owner wants the architect to transfer ownership of the architectural plans and specifications to the owner as a prerequisite to granting the design contract. What is the best solution if the architect wants to get the job while still protecting the interests of her firm? **Check the two that apply.**
    a. Tell the owner that the architect should have copyrights of all documents generated by the architect per the AIA documents.
    b. Transfer ownership of the architect's plans and specifications to the owner and require the owner to sign a waiver releasing the architect of the liabilities for unauthorized use of the documents.
    c. Negotiate with the owner and try to seek joint-ownership of the plans and specifications.
    d. Refuse the owner's request.

28. Which of the following is characteristic of a fast-track project? **Check the two that apply.**
    a. The design phase occurs before the construction phase.
    b. The design phase overlaps the construction phase.
    c. Multiple bid packages are involved.
    d. One bid package is utilized to simplify the construction process.

29. A client asks an architect to design an office building. The architect should advise the client that a detailed program is likely to result in (**Check the two that apply.**)
    a. lower construction cost
    b. fewer change orders
    c. a more aesthetically pleasing building
    d. extended construction time

30. SPL, as applied to buildings, represents the
    a. sound pressure level
    b. sound power level
    c. solar power level
    d. security protection level

31. Which of the following is the most important consideration in an architect/owner contract? **Check the two that apply.**
    a. scope of services
    b. consultants
    c. type of construction
    d. architectural service fees

32. Which of the following is typically not required for a new library building plan check?
    a. a planning department plan check fee
    b. a building department plan check fee
    c. a fire department plan check fee
    d. a health department plan check fee
    e. a school district fee
    f. a drainage fee

33. According to AIA document B101, programming is
    a. part of the design process
    b. standard practice
    c. not a basic service
    d. reimbursable

34. The cost for a geotechnical survey is typically borne by the
    a. owner
    b. architect
    c. contractor
    d. federal government

35. The best way to reduce the number of change orders is to
    a. hold regular coordination meetings
    b. use an outside peer review service
    c. issue multiple bid packages
    d. have owners review the plans
    e. finish coordination and quality control before issuing the bid package

36. Which of the following is an effective way to improve the quality of construction documents? **Check the two that apply.**
    a. communication among employees
    b. hiring qualified employees
    c. having owners review the plans
    d. using outside consultants

37. The size and number of parking stalls is typically regulated by which of the following? **Check the two that apply.**
    a. deed restrictions
    b. building codes
    c. zoning ordinances
    d. life-safety codes

38. An architect is working on a project with her MEP consultants. The owner asks the electrical engineer to assist him in adjusting some of the utility easements. What should the electrical engineer do?
    a. Provide the service without extra charge since it is part of the basic service.
    b. Send the owner an additional service request and obtain the owner's authorization before proceeding.
    c. Send the architect an additional service request to obtain the architect's authorization before proceeding.
    d. Send the architect an additional service request, and wait until the architect forwards this request to the owner and obtains the owner's authorization before proceeding.

39. AIA Document C401-2007 includes a new flow-down provision to extend the responsibilities and rights between which of the following parties?
    a. the owner and the contractor
    b. the architect and the owner
    c. the architect and the contractor
    d. none of the above

40. Outline specifications during the schematic design stage are typically broken down by
    a. divisions
    b. disciplines
    c. costs
    d. the critical path method

41. Bid alternates to choose between stone veneer and brick veneer, double-glazing and single-glazing, and clay roof tiles and asphalt shingles are most likely the architect's attempt to
    a. incorporate environmental friendly options
    b. control construction costs
    c. anticipate HOA CC&R requirements
    d. address neighborhood concerns

42. Which of the following design features encourages people to stay in a public plaza?
    a. beautiful pavement patterns
    b. trees with large canopies
    c. elegant benches
    d. fountains

43. Soils reports are typically paid for by the?
    a. city
    b. civil engineer
    c. architect
    d. structural engineer
    e. owner
    f. contractor

44. A specific plan is typically developed or paid for by the?
    a. city
    b. civil engineer
    c. architect
    d. structural engineer
    e. owner
    f. contractor

45. An EIR is typically paid for by the?
    a. city
    b. civil engineer
    c. architect
    d. structural engineer
    e. owner
    f. contractor

46. During the construction documents phase of work, the architect receives the plan check corrections. Which of the following is the most effective way to coordinate the engineering consultant's work?
    a. Send the entire plan check corrections list to all the consultants, and then start to review the plan check corrections list.
    b. Review the plan check corrections list first, and then send only the relevant part of the plan check corrections list to the related consultants.
    c. Review and mark up the plan check corrections, mark up the consultants' plans per the list, and then send only the relevant part of the plan check corrections list and the marked-up sheets of the consultants' plans to the related consultants.
    d. Review and mark up the plan check corrections, mark up the consultants' plans per the list, and then send only the relevant part of the plan check corrections list and entire set of the consultants' plans with the mark-ups to the related consultants.

47. The purpose of the programming process is to establish (**Check the two that apply.**)
    a. the cost limit
    b. evaluation of materials
    c. the OPR
    d. the BOD

48. Which of the following is true according to A101–2007, Standard Form of Agreement Between Owner and Contractor where the basis of payment is a Stipulated Sum (CCDC Document 2)? **Check the two that apply.**
    a. Mediation is binding in most states.
    b. Arbitration is binding in most states.
    c. Mediation is mandatory.
    d. Arbitration is submitted to AIA.

49. The purpose of zoning ordinances, include all of the following EXCEPT (**Check the two that apply.)**
    a. to preserve the "character" of a community
    b. to limit building heights
    c. to control construction cost
    d. to reduce seismic impact

50. Buildings in urban areas of American cities may be allowed to exceed FAR and height limits required by zoning because
    a. the developer provides a public park on the site
    b. the developer uses green materials
    c. the developer provides more efficient vertical transportation to the building
    d. the developer uses more materials approved by the city

51. Which of the following can meet a city's ponding requirements? **Check the two that apply.**
    a. a decorative landscape pond
    b. an area where rainwater can be retained and discharged into a storm drain
    c. a trench drain system
    d. an underground tank to collect and use rainwater

52. Which of the following are not typical city zoning requirements? **Check the two that apply.**
    a. life safety issues
    b. coverage ratio
    c. FAR
    d. energy efficiency of the building envelope

53. An architectural program should not include which of the following? **Check the two that apply.**
    a. a budget
    b. a detailed list of doors and their related sizes
    c. design guidelines for footings
    d. a list of required spaces and their relationships

54. Governing agencies regulate the development of projects through which of the following? **Check the two that apply.**
    a. General plans
    b. Specific plans
    c. CC&R
    d. plumbing permits

55. The area of a space is determined by which of the following? **Check the three that apply.**
    a. number of people utilizing the space
    b. equipment to be placed in the space
    c. zoning codes
    d. activities within the space

56. The total gross area of a building is equal to
    a. the sum of the net area and the building wall thickness
    b. the sum of the net area and the assigned area
    c. the sum of the net area and the secondary area
    d. none of the above

57. Which of the following will affect a project's budget? **Check the two that apply.**
    a. location of the project
    b. inflation
    c. schedule of the city's architectural review board
    d. traffic conditions near the site

58. The cost index for City A is 1268, and the cost index for City B is 1663. If a building's construction cost is 3.3 million dollars in City A, how much will this building cost in City B?
    a. 3.25 million dollars
    b. 4.33 million dollars
    c. 5.23 million dollars
    d. 6.38 million dollars

59. Which of the following will affect a project's schedule? **Check the three that apply.**
    a. the client
    b. the number of project team members
    c. the architect
    d. insurance

60. Which of the following will restrict a project's development? **Check the three that apply.**
    a. the municipal codes
    b. the building codes
    c. the experience of construction workers
    d. the fire department

61. Which of the following are major considerations during programming? **Check the four that apply.**
    a. function
    b. form
    c. time
    d. economy
    e. neighborhood
    f. green building features

62. Which of the following does a soils report typically include? **Check the two that apply.**
    a. landscaping
    b. seismic considerations
    c. utilities
    d. footing and foundation design guidelines

63. Which of the following is the correct order of arranging the units used in the Public Land Survey System (PLSS) in the US, from large to small?
    a. check, section, township
    b. section, township, check
    c. check, township, section
    d. section, check, township

64. Which of the following is a measure to improve water quality? **Check the two that apply.**
    a. retention pond
    b. detention pond
    c. concrete swale
    d. trench drain

65. Which of the following is the best method to reduce asbestos exposure in an existing building?
    a. Remove the asbestos.
    b. Keep asbestos-containing materials in place without disturbing it.
    c. Seal off the spaces with asbestos-containing materials.
    d. Ban children from rooms with asbestos containing materials.

66. An architect is laying out the parking stalls along a curb. Which of the following parking configurations will accommodate the most cars per 100 linear feet of curb?
    a. 90 degree parking
    b. 60 degree parking
    c. 45 degree parking
    d. 30 degree parking

67. Which of the following does not regulate a building's height?
    a. zoning codes
    b. building codes
    c. CC&R
    d. ADA

68. Which of the following do not need to comply with ADA Accessibility Guidelines (ADAAG)?
    a. a city hall building
    b. a single family home on a private property
    c. a construction trailer
    d. both b & c

69. In a construction project, a contractor's insurance will pay for claims from the owner, if the owner waives his rights to sue for and recover from the contractor. This arrangement is an example of (**Check the two that apply.**)
    a. a waiver of subrogation
    b. an exclusive right
    c. an exculpatory clause
    d. a waiver of abrogation

70. Per AIA Document B101-2007, which of the following regarding mediation is true? **Check the two that apply.**
    a. Mediation shall always precede litigation.
    b. The prevailing party will be reimbursed for the mediation fees.
    c. Arbitration shall always precede mediation.
    d. The parties shall share the mediation fees equally.

71. Per AIA Document B101-2007, who should furnish the program for a project?
    a. owner
    b. architect
    c. contractor
    d. building official

72. Which of the following can use eminent domain to acquire land for the project development? **Check the two that apply.**
    a. a public school
    b. a toll road
    c. an interstate freeway
    d. a shopping center

73. Which of the following project types receives the most tax incentive with regard to historical buildings?
   a. preservation
   b. rehabilitation
   c. restoration
   d. reconstruction

74. Which of the following is the correct order to arrange building types based on their building efficiencies, from high to low?
   a. department store, office, apartment, hospital
   b. department store, apartment, office, hospital
   c. office, apartment, hospital, department store
   d. office, apartment, department store, hospital

75. Which of the following are not considered laws?
   a. USGBC LEED reference guides
   b. building codes
   c. ADA
   d. municipal codes
   e. EPA Codes of Federal Regulations

76. Which of the following can reduce stormwater runoff and alleviate the urban heat island effect? **Check the three that apply.**
   a. increasing the site coverage ratio
   b. increasing Floor Area Ratio (FAR)
   c. using a vegetated roof
   d. using porous pavement with high albedo
   e. building a retention pond on the site

77. Recycled materials will contribute to which of the following?
   a. traffic alleviation and smog reduction
   b. protection of virgin materials
   c. energy savings
   d. MEP cost savings

78. According to the USGBC, which of the following is not graywater?
   a. water from kitchen sinks
   b. water from toilet
   c. harvest rainwater
   d. water from outdoor area drains
   e. none of above
   f. a, b, c, and d

79. Which of the following is not true? **Check the two that apply.**
    a. Water from kitchen sinks can be reused for landscape irrigation or flushing toilets.
    b. Water from kitchen sinks cannot be reused for landscape irrigation or flushing toilets.
    c. Reclaimed water requires special piping with a different color.
    d. Reclaimed water cannot reduce potable water use.

80. Which of the following are must-have features of open spaces? **Check the two that apply.**
    a. vegetation
    b. shade
    c. brownfields
    d. pervious Areas
    e. hardscape Areas

81. A LEED certified building has the following extra costs when compared with a conventional building?
    a. hard costs
    b. soft costs
    c. storm control costs
    d. life cycle analysis costs
    e. life cycle cost analysis

82. All LEED rating systems include credits for:
    a. emission measurement.
    b. radon alleviation.
    c. minimum energy performance.
    d. innovation.

83. Which of the following have the lowest ODP?
    a. CFCs
    b. HCFCs
    c. HFCs
    d. This is hard to determine.

84. Choose the post-consumer item from the following.
    a. construction debris sent to a recycle facility
    b. scraps from a manufacturing process
    c. books from a print overrun
    d. scraps from a manufacturing process that were reclaimed and used in a different manufacturing process

85. Which of the following evaluates the environmental performance of services and products?
    a. ASTM
    b. ISO 14000
    c. ANSI
    d. LEED

B. Mock Exam: Site Zoning Vignette

1. Directions and code

The directions and **code** are the same as the Site Zoning sample vignette in the official NCARB exam guide. The NCARB **directions** and **code** has been very consistent throughout various versions of the ARE.

You can download the official exam guide and practice program for PPP division at the following link:
http://www.ncarb.org/en/ARE/Preparing-for-the-ARE.aspx

2. Program
The program is the same as the Site Zoning sample vignette in the official NCARB exam guide, except the following (figure 3.2):
1) Surface improvements are prohibited within 6 ft of any property line.
2) Construction of buildings is prohibited within the following setbacks. (All setbacks are measured from the property lines of the two new lots.)
3) Front yard setbacks shall be considered only from Main Street.
   - Front yard setbacks from property line along Main Street: 20 ft
   - Rear yard setbacks: 25 ft
   - Side yard setbacks: 10 ft
4) Construction of buildings and other surface improvements is prohibited within 20 ft of the lake high water line.
5) Construction of buildings is prohibited within the existing drainage easement.
6) The maximum building height limit within 55 ft of the east property line in Lot B shall be 50 ft above the benchmark elevation.
7) The maximum building height limit between 0 ft and 40 ft of the west property line of Lot A shall be 25 ft above the grade at the property line.
8) Maximum building height limit shall be 80 ft above the benchmark elevation.
9) The maximum building envelope is restricted to an elevation defined by a 135-degree line rising westward from a point at an elevation of 25 ft directly above the benchmark.

Note:
For your convenience, we have placed the DWG file for figure 3.2 on our website at:
http://GreenExamEducation.com

Here are the simple steps for you to download the DWG file for FREE:
- *Go to http://GreenExamEducation.com/*
- *Click on "Free Downloads and Forums" on the top menu.*
- *Follow the instructions on page 59.*
- *After you download the DWG file, you can install the DWG file to use with the NCARB ARE 4.0 software by following our instructions.*

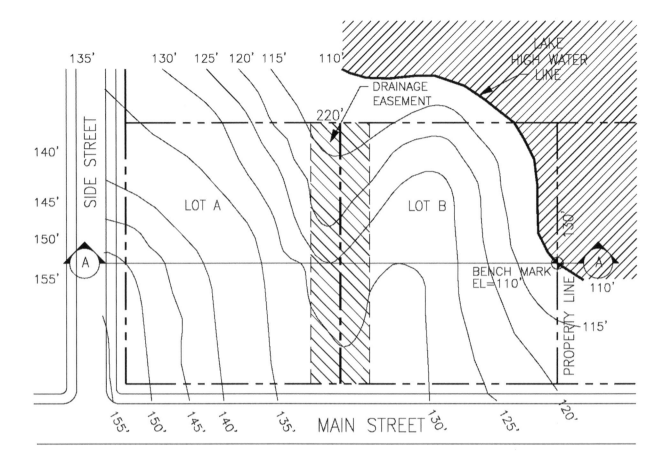

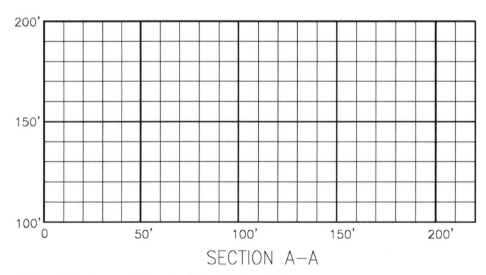

SECTION A-A

Figure 3.2   Background for the Site Zoning Vignette

C. Instructions on Installing Alternate DWG Files for Use with NCARB Software

1) **Right click** the **Start** button on the lower left-hand corner of your computer to open your **Windows Explore** (figure 3.3).
2) Go to the folder where you placed the downloaded DWG file. On the top pull-down menu, under **View**, select **Details**. You should see an extension for all the file names, i.e., a dot (.) followed by three letters. For example, the AutoCAD file name for our alternate drawing is **PPP Figure 3.2.dwg**; the ".dwg" is the extension (figure 3.4).
3) If you do NOT see an extension for all the file names, see the following instructions. (These directions are for Windows Vista and Window 7, but Windows XP is similar.)
   - **Windows Explore > Organize > Folder and Search Options** (figure 3.5) A menu window will pop up. In that menu, select **View** (figure 3.6).
   - You will see a list with several boxes checked. Scroll down and **uncheck** the box for **Hide extensions for known file types.**
   - Select **Apply to Folders** and a menu window will pop up. Select **Yes** to get out of the **View** menu (figure 3.7).
4) **Windows Explore > Computer > C: Drive > Program Files > NCARB.** Select the folder for the NCARB ARE division that you are working on (figure 3.8). For example, **C:\Program Files\NCARB\Programming, Planning, and Practice**
5) Open the folder and you will see files ending in .AUT and .DWG. The .AUT files are the program information and .DWG files are the templates for the practice vignettes. This is important.
6) Make a new folder called **Backup** under the **Programming, Planning, and Practice** folder. Copy all the .AUT files and .DWG files to the **Backup** folder. For PPP, we only have one DWG format file: **B6TUT3W1.dwg.**
7) Make a new subfolder called **Alt** in the **Programming, Planning, and Practice** folder. Copy the alternate DWG file(s) (**PPP Figure 3.2.dwg**) that you want to use into the Alt folder. Rename the alternate DWG file(s) to match the name of the original NCARB DWG files. For our example, we will rename **PPP Figure 3.2.dwg** as **B6TUT3W1.dwg** to match the original NCARB DWG file.
8) Copy the alternate DWG file(s) to the NCARB folder for your vignette (**C:\Program Files\NCARB\Programming, Planning, and Practice**) and overwrite the original NCARB DWG files (figure 3.9).

Note: NCARB practice software ARE 4.0 ONLY works with AutoCAD Release 12 version. If you have a DWG file that is in an AutoCAD Release 13 version or higher, the NCARB practice software ARE 4.0 will NOT work, and you have to save the DWG file as AutoCAD Release 12 version file.

When you save the DWG file AutoCAD Release 12 version file, you may lose some information such as the pen weights, the leader arrow size, etc. However, you can still read the plan and understand the concepts for this exercise.

9) Open the NCARB practice software. You may get an error message that says the DWG has been changed. Just ignore it and click OK.
10) Now you can start to work on your solution using the NCARB software.

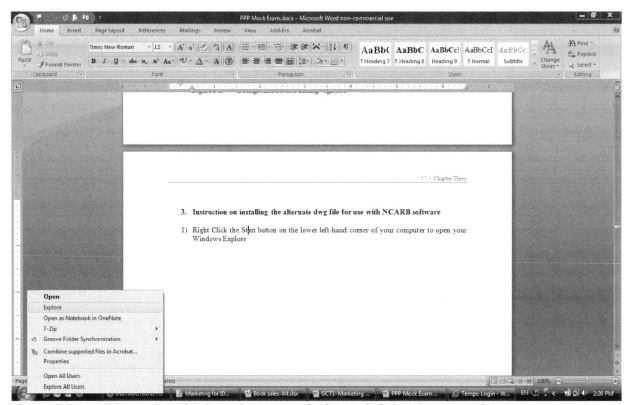

Figure 3.3   **Right click** the **Start** button on the lower left-hand corner of your computer to open **Windows Explore.**

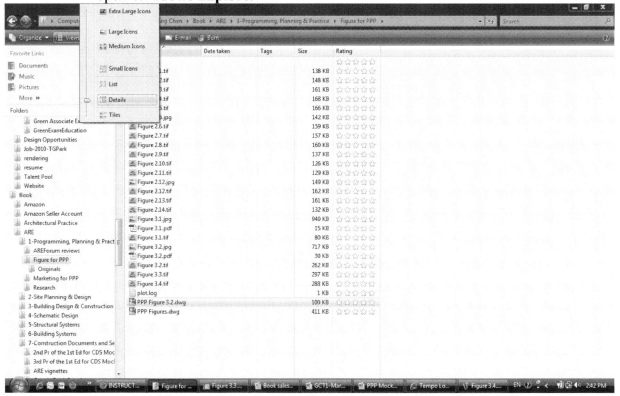

Figure 3.4   You should see an extension for all the file names, i.e., a dot (.) followed by three letters.

Figure 3.5   **Windows Explore > Organize > Folder and Search Options**

Figure 3.6   A menu window will pop up. In that menu, select **View.**

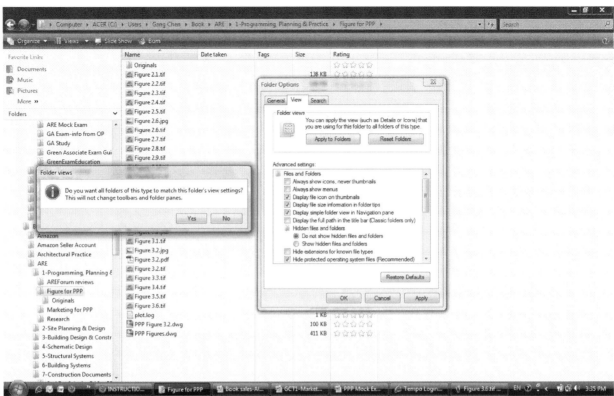

Figure 3.7   Select **Apply to Folders**. and a menu window will pop up. Select **Yes** to get out of the **View** menu.

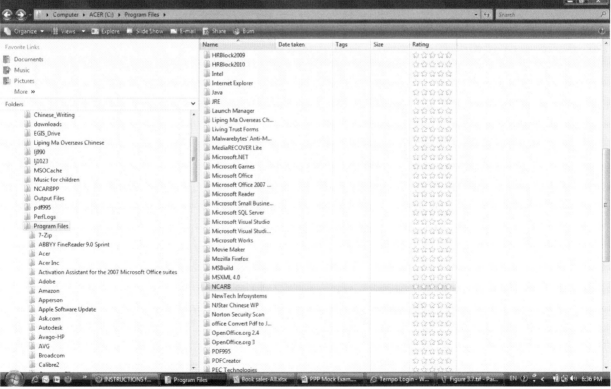

Figure 3.8   **Windows Explore > Computer > C: Drive > Program Files > NCARB.** Select the folder for the NCARB ARE division that you are working on.

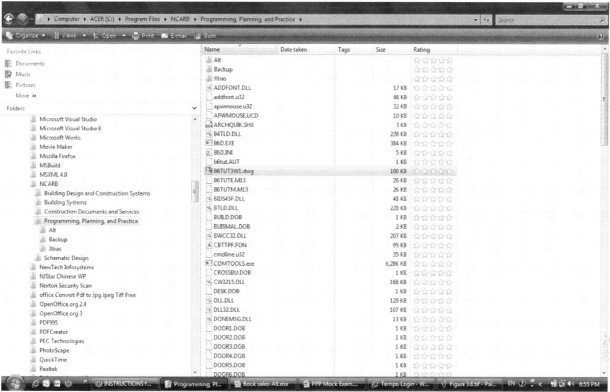

Figure 3.9   Copy the alternate DWG files to the NCARB folder for your vignette and overwrite the original NCARB DWG files.

Chapter Four

ARE Mock Exam Solutions for
Programming, Planning & Practice (PPP) Division

A. Mock Exam Answers and Explanations: PPP Multiple-Choice (MC) Section

Note: If you answer 60% of the questions correctly, you pass the MC Section of the exam.

1. Answer: b and c

    The following are true about lead-based paints in existing U.S. buildings:
    - Lead-based paints have been banned in all new U.S. residential construction.
    - Lead-based paints pose a greater health risk to children seven years or younger.

    The following are NOT true about lead-based paints in existing U.S. buildings:
    - Lead-based paints have been banned in all U.S. buildings. (Lead-based paints in existing homes are still acceptable.)
    - Lead-based paints pose a greater health risk to the elderly.

2. Answer: d and e
    The following consultants are more likely to be involved with the preliminary design of a project:
    - landscape architect (preliminary site plan and master planning, etc.)
    - civil engineer (site contours and preliminary grading plans, etc.)

    The following consultants are less likely to be involved with the preliminary design of a project:
    - mechanical engineer
    - electrical engineer
    - plumbing engineer

3. Answer: d
    Poor ventilation generates the most indoor air quality problems.
    The following are incorrect answers:
    - lack of air conditioning equipment (Natural ventilation can alleviate indoor air quality problems.)
    - poor construction practice (This only applies to buildings under construction.)
    - soils brought in by pedestrian traffic (This is not a major source of indoor air quality problems.)
    - no natural ventilation (Artificial ventilation can alleviate indoor air quality problems.)

4. Answer: a

A multiple prime is the correct answer. A **multiple prime** is the arrangement in which the owner contracts directly with several contractors instead of a single prime contractor.

The following are the incorrect answers and their definitions:
- **an associated firm:** a firm controlled by another firm to an extent that is less than a subsidiary
- **a joint venture**: for a finite time, two firms join together to develop a new entity and new assets by contributing equity
- **partnering**: to associate as partners

5. Answer: c

Completing a feasibility study is the best choice.

The other choices listed are good suggestions, but they are NOT the best.
- compare the cost of the two choices
- seek LEED certification for the home
- complete a life safety study

6. Answer: c

EPA identified maximum sound level to protect against hearing loss and other disruptive effects from noise, such as sleep disturbance, stress, and learning detriment as 70 db.

See the following link:
http://en.wikipedia.org/wiki/Sound_pressure

You should become familiar with some common sound levels.

7. Answer: b

Blocking and stacking are terms used in programming. **Stacking** is an activity of programming, in which floors or areas of floors are assigned to departments based upon their adjacency and support requirements.

Blocking is an activity of programming, in which <u>departments are assigned</u> to a particular area of a floor based upon adjacency and support requirements.

The following are incorrect answers:
- terms used in masonry construction
- terms used in structural calculations
- terms used in design development

8. Answer: a

   Unit-area cost is the most common method of estimating construction cost at the schematic design stage.

   Time and materials is for a task without a fixed scope. **Fixed fee** can be based on several estimating methods. **Value engineering** is a nice name for construction cost reduction.

9. Answer: b

   The first C in CC&R stands for Covenants. **CC&R is Covenants, Conditions and Restrictions**. They are limitations and rules set by developer, a builder, neighborhood association, and/or homeowner association. All townhomes, condos, as well as most planned unit developments, and established neighborhoods have CC&Rs.

10. Answer: a

    Square-foot [square-meter] costs of similar buildings is a common method of figuring construction cost estimates at the programming stage.

    The following are incorrect answers:
    • the design development stage
    • the construction document stage
    • bidding
    Detailed estimates are possible at these stages.

11. Answer: a and d

    An operating pro-forma is
    • a year-by-year projection of a project's expenses and income
    • of particular interest to a project's lenders and investors

    The following are incorrect answers:
    • a detailed breakdown of a project's operation expense
    • a return on investment ratio

12. Answer: a, b and d

    The following can cause mold inside a building wall:
    • poor ventilation
    • poor drainage
    • organic feedstock

    The following are incorrect answers:
    • flashing (Lack of proper flashing can cause mold.)
    • EIFS (Lack of proper flashing and drainage can cause mold in a wall with an EIFS system, although not in EIFS itself.)

13. Answer: a, b, e, and f
An architectural project program should include the following:
- a basis of design (BOD)
- owner project requirements (OPR)
- a budget
- type and quantity of spaces

The followings are incorrect answers:
- type of structural system (This should be part of design development decisions.)
- type of HVAC system (This should be part of design development decisions.)

14. Answer: b
The project schedule in figure 3.1 is known as a critical path method (CPM).

A Gantt chart is a type of bar chart used frequently in project schedules. **Program evaluation and review technique (PERT)** is a statistical tool, used in project management. **Project cycle method (PCM)** or **Project cycle management method (PCM)** is a tool to manage the whole project cycle.

15. Answer: d
The total numbers of days needed to finish the project per the project schedule in figure 3.1 is 38. The total number of days is determined by the critical path, which is shown as bolded arrows in figure 3.1. Each solid arrow represents an activity. The numbered circles are beginning and/or end point of an activity. *All activities leading to a circle must be finished before you can go to the next step.* You need to pick the *worst* case leading to each circle to calculate the time needed to finish an activity.

16. Answer: c
The dashed arrows in figure 3.1 are known as dummies. **Dummies** indicate relationship. They are not activities and have no duration.

The following are incorrect answers or **distracters**:
- knots
- paths
- processes

17. Answer: d and e
The following are not a part of architectural programming, and therefore the correct answers:
- codes and regulations
- geotechnical surveys

The following are part of architectural programming
- functional and operational requirements
- scoping
- decision-making processes

18. Answer: a

The programming stage is the best time to make changes to a project. The earlier the changes are made, the lesser the cost impact, and the better the opportunities for influence.

See the following link:
http://www.wbdg.org/design/dd_archprogramming.php

19. Answer: a, c, d, and e

The following people should be invited to the project programming meeting:

- the owners
- the tenants
- the design consultants
- the leasing team

The building officials could be invited, but they are NOT as important as other parties listed above during this stage.

The neighbors should not be invited to the meeting.

20. Answer: b

Utility availability can have an <u>unobvious</u> high cost impact at a given site. Many people may neglect this factor when selecting a site.

The following factors have less cost impact at a <u>given</u> site:

- topography (This is an obvious factor if it has high cost impact.)
- traffic
- vegetation and wildlife
- zoning requirements

See the following link:
http://www.wbdg.org/design/dd_archprogramming.php

21. Answer: **Building efficiency** is <u>the ratio of areas that are assigned to a function (net assignable square feet or NASF) to gross square feet (GSF)</u>.

Building efficiency = NASF/GSF

See the following link:
http://www.wbdg.org/design/dd_archprogramming.php

22. Answer: Tare space is <u>the area needed for mechanical, electrical and telephone equipment, walls, wall thickness, circulation, and public toilets.</u>

   The total building area includes the NASF and tare areas.

23. Answer: b, c and d
   The following are essential elements for architectural programming:
   - project type
   - goals and objectives
   - quantitative requirements

   The following are nonessential elements for architectural programming:
   - topography
   - vegetation and wildlife
   - zoning requirements

24. Answer: b
   After the award of a construction contract, the contractor shall submit a schedule of construction for architect review.

   The following are incorrect answers or **distracters**:
   - a punch list (The contractor shall submit this at the end of construction.)
   - a list of potential change order (The contractor does NOT have to submit this.)
   - a change order log (The contractor does NOT have to submit this.)
   - a submittal log (The contractor does NOT have to submit this.)

25. Answer: a
   According to the Americans with Disabilities Act (ADA) [CSA-B651-04] Accessibility Guidelines, minimum clear passage width for single wheelchair is 36 inches (915 mm) minimum along an accessible route, but may be reduced to 32 inches (815 mm) minimum at a point for a maximum depth of 24 inches (610 mm).

   See the following link:
   http://www.access-board.gov/adaag/html/figures/fig1.html

26. Answer: a
   Unit price is the most accurate type of estimate.

   The following are incorrect answers or **distracters**:
   - An estimate based on the construction cost of similar buildings
   - cost of building systems
   - historical data of the same type of construction

27. Answer: b and c

The best solutions for the architect to get the job while still protecting the interests of her firm are:

- Transfer ownership of the architectural plans and specifications to the owner and require the owner to sign a waiver releasing the architect of the liabilities for unauthorized use of the documents.
- Negotiate with the owner and try to seek joint-ownership of the plans and specifications.

Since the owner wants the architect to transfer ownership of the architectural plans and specifications to the owner as a prerequisite to granting the design contract, the architect is unlikely to get the job if she chooses the following two options:

- Tell the owner that the architect should have copyrights of all documents generated by the architect per the AIA documents.
- Refuse the owner's request.

28. Answer: b and c

The following are characteristic of fast-track projects:

- The design phase overlaps the construction phase; construction often starts before the plans are complete.
- Multiple bid packages are involved, such as the foundation bid package, the superstructure bid package, and the exterior enclosure bid package.

The following are incorrect answers or **distracters**:

- The design phase occurs before the construction phase.
- One bid package is utilized to simplify the construction process.

29. Answer: a and b

The architect should advise the client that a detailed program is likely to result in:

- lower construction cost
- fewer change orders

The following are incorrect answers:

- a more aesthetically pleasing building
- extended construction time

30. Answer: a

SPL, as applied to buildings, represents the sound pressure level. It can be thought of as the effect of sound from a source in an enclosed space.

The following are incorrect answers:

- sound power level (It should be PWL)
- solar power level
- security protection level

31. Answer: a and d
    The following are the most important considerations in an architect/owner contract:
    - scope of services
    - architectural service fees

    The following may NOT even be discussed in an architect/owner contract: (They are the incorrect answers.)
    - consultants
    - type of construction

32. Answer: d
    A health department plan check fee is typically not required for a new library building because a library does NOT sell food. Health department plan check fees are required for buildings that have food processing departments or sell foods, including pre-packaged food such as candy.

    The following fees are typically required for a new library building plan check as well as on most other new buildings:
    - a planning department plan check fee
    - a building department plan check fee
    - a fire department plan check fee
    - a school district fee
    - a drainage fee

33. Answer: c
    According to section 4.11 of AIA Document B101, Standard Form of Agreement Between Owner and Architect, programming is not a basic service, but an additional services instead.

    The following are incorrect answers:
    - part of the design process
    - standard practice
    - reimbursable (for telephone fees, printing, etc. An additional service is NOT called reimbursable.)

34. Answer: a
    The cost for a geotechnical survey is typically borne by the owner.

    The following are incorrect answers:
    - architect
    - contractor
    - federal government

35. Answer: e

The best way to reduce the number of change orders is to finish coordination and quality control before issuing the bid package.

Issuing multiple bid packages will NOT reduce the number of change orders.

The following will reduce the number of change orders, but they are NOT the best way, and therefore the incorrect answers:
- hold regular coordination meetings
- use an outside peer review service
- have owners review the plans

36. Answer: a and b

Hiring qualified employees and communication among employees are effective ways to improve the quality of construction documents.

The following are helpful, but they are NOT the best way, and therefore the incorrect answers:
- have owners review the plans
- use outside consultants

37. Answer: b and c

The size and number of parking stalls is typically regulated by:
- building codes (In some states, like California, the California Building Code dictates the number of handicap parking stalls.)
- zoning ordinances (most cities)

The following are incorrect answers:
- deed restrictions (These restrictions control easements, CC&R, etc., but NOT the size and number of parking stalls.)
- life-safety codes (These codes control the stair and exit widths, etc., but NOT the size and number of parking stalls.)

38. Answer: c

The electrical engineer should send the architect an additional service request, and obtain the architect's authorization before proceeding. This is because normally the electrical engineer does not have a direct contract with the owner, and is simply a sub-consultant under the architect's master design contract.

If the architect is smart, she would create a related additional service request, forward the request to the owner, and obtain the owner's consent before authorizing the electrical engineer to proceed. However, the architect does have the option of authorizing the electrical engineer to proceed WITHOUT obtaining the owner's authorization first. In this case, the architect is taking on a risk herself.

The following are incorrect answers:
- Provide the service without extra charge since it is part of the basic service.
- Send the owner an additional service request and obtain the owner's authorization before proceeding.
- Send the architect an additional service request, and wait until the architect forwards the request to the owner and obtains the owner's authorization before proceeding.

We do beta tests of our mock exam questions before their final release in order to improve the quality of our books, and make sure that our readers get the best products. The following are questions we received from a reader regarding this question and our related responses. We add them below because we think you will benefit from them.

Question: First, what business does the owner have talking to the electrical engineer about easements?
Response: This is based on a real world situation that occurred two months ago in my practice.
Question: Second, what about the authorities that established the easement in the first place? Perhaps clarifying the type of easement would help as well. Shouldn't they be approached? And how?
Response: Good point. It was the electrical, cable, and telephone company utility easements. The owner was trying to hire the electrical engineer directly to deal with these firms.
Question: Third, does the owner ever have a say in an easement on their property placed by the electrical company? (Which I am assuming is what is being questioned.) I had the impression that these types of easements (assuming it is the electrical power service) are permanent unless the company changes it themselves.
Response: If the easements are no longer needed, then the owner can apply and request the utility firms to abandon these easements,
Question: In the end, according to your explanation, shouldn't the answer be c and d?
Response: No, ONLY c is the correct answer because the architect does have the option of authorizing the electrical engineer to proceed WITHOUT obtaining the owner's authorization first. In this case, the architect is taking on a risk herself. The electrical engineer works for the architect, NOT the owner.

39. Answer: b
The AIA Document C401-2007 includes a new flow-down provision that extends the responsibilities and rights contracted between the architect and owner, down to the agreement made between the consultants and the architect.

The following are incorrect answers:
- the owner and the contractor
- the architect and the contractor
- none of the above

40. Answer: a
Outline specifications during the schematic design stage are typically broken down into divisions.

The following are incorrect answers:
• disciplines
• costs
• the critical path method

41. Answer: b
Bid alternates to choose between stone veneer and brick veneer, double-glazing and single-glazing, and clay roof tiles and asphalt shingles are most likely the architect's attempt to control construction costs.

The following are incorrect answers:
• incorporate environmental friendly options
• anticipate HOA CC&R requirements (HOA stands for Home Owner's Association; CC&R stands for Covenants, Conditions and Restrictions.)
• address neighborhood concerns

42. Answer: c
Elegant benches encourage people to stay in a public plaza. Research has shown that people tend to stay longer at places where they can sit down.

The following are good features, but they are NOT the best answer:
• beautiful pavement patterns
• trees with large canopies
• fountains

43. Answer: e
Soils reports are typically paid for by the owner.

The following are incorrect answers:
• city
• civil engineer
• architect
• structural engineer
• contractor

44. Answer: a
A Specific Plan is typically developed or paid for by the city.

The City of San Marcos, California has a very good definition of a General Plan and a Specific Plan. This definition also applies to other cities.

"The **General Plan** is the long-term policy guide for the physical, economic, and environmental growth of a city and represents the community's vision of its ultimate physical growth."

"A **Specific Plan** is a comprehensive planning document that guides the development of a defined geographic area in a mix of uses including residential, commercial, industrial, schools, parks, and open space."

See the following link:
http://www.ci.san-marcos.ca.us/index.aspx?page=323

The following are incorrect answers:
- civil engineer
- architect
- structural engineer
- owner
- contractor

45. Answer: e
EIR stands for Environmental Impact Report. The owner typically pays for an EIR.

The following are incorrect answers:
- city
- civil engineer
- architect
- structural engineer
- contractor

46. Answer: c
During the construction documents phase of work, the architect receives the plan check corrections. The following is the most effective way to coordinate the engineering consultant's work:
- Review and mark up the plan check corrections, mark up the consultants' plans per the list, and then send only the relevant part of the plan check corrections list and the marked-up sheets of the consultants' plans to the related consultants.

The following are incorrect answers:
- Send the entire plan check corrections list to all the consultants, and then start to review the plan check corrections list. *Note: This is NOT efficient because you are asking the consultants to deal with the same thing twice.*
- Review the plan check corrections list first, and then send only the relevant part of the plan check corrections list to the related consultants. *Note: This is NOT efficient because you will have to send the marked-up sheets of the consultants' plans to the related consultants later.*
- Review and mark up the plan check corrections, mark up the consultants' plans per the list, and then send only the relevant part of the plan check corrections list and entire set

of the consultants' plans with the mark-ups to the related consultants. ***Note: This is NOT efficient because the consultants should already have a complete set of plans for coordination, and you do NOT need to re-send the entire set of the consultants' plans with the mark-ups.***

47. Answer: c and d
The purpose of the programming process is to establish
- the OPR (the Owner's Project Requirements)
- the BOD (Basis of Design)

The following are incorrect answers:
- the cost limit
- evaluation of materials

48. Answer: b and c
The following items are true according to A101–2007, Standard Form of Agreement Between Owner and Contractor where the basis of payment is a Stipulated Sum (CCDC Document 2):
- Arbitration is binding in most states.
- Mediation is mandatory.

Mediation is NOT binding.

Arbitration is NOT handled by AIA, but instead by the American Arbitration Association (AAA).

49. Answer: c and d
Pay attention to the word "EXCEPT."

The purpose of zoning ordinances includes the following, and therefore they are the incorrect answers:
- to preserve the "character" of a community
- to limit building heights

The purpose of zoning ordinances does NOT include the following, and therefore they are the correct answers:
- to control construction cost
- to reduce seismic impact (This is the purpose of the building codes, NOT the purpose of the zoning ordinances.)

50. Answer: a
Buildings in urban areas of American cities may be allowed to exceed FAR and height limits required by zoning because the developer provides a public park on the site. A public park is considered a public amenity. **FAR** is floor area ratio, and is the ratio of the total building area to the site area.

The following are nice design features but they are NOT considered <u>public</u> amenities, and are unlikely to achieve incentives such as exceeding FAR and height limits required by zoning. Therefore, they are incorrect answers.

* The developer uses green materials.
* The developer provides more efficient vertical transportation to the building.
* The developer uses more materials approved by the city.

51. Answer: a and b

The following can meet a city's ponding requirements:

* a decorative landscape pond
* an area where rainwater can be retained and discharged into a storm drain

The following are incorrect answers:

* a trench drain system
* an underground tank to collect and use rainwater (Because the tank is underground, the rainwater is not considered ponding.)

52. Answer: a and d

Pay attention to the word "not."

The following are typically not city zoning requirements, and therefore the correct answers:

* life safety issues
* energy efficiency of the building envelope

The following are typically city zoning requirements, and therefore the incorrect answers:

* **coverage ratio** (This is the ratio of the <u>first</u> floor area to the site area.)
* FAR

53. Answer: b and c

Pay attention to the word "not."

An architectural program should not include the following, and therefore they are the correct answers:

* a detailed list of doors and their related sizes (This is typically part of the construction documents, and not an architectural program.)
* design guidelines for footings (These are typically part of the soils report prepared by a soils engineer.)

An architectural program should include the following:

* a budget
* a list of required spaces and their relationships

54. Answer: a and b
Governing agencies regulate the development of projects through:
- General plans
- Specific plans

See answer 44 on pages 75 and 76 for definitions of these terms as well as a website link for further information.

The following are incorrect answers:
- CC&R (This is used by Home Owners' Associations, or HOA.)
- plumbing permits (These are used to regulate specific buildings, and are NOT used for the development of projects.)

55. Answer: a, b and d
The area of a space is determined by the following:
- number of people utilizing the space
- equipment to be placed in the space
- activities within the space

Zoning codes is an incorrect answer because zoning codes regulate the overall development of a building, not the individual spaces.

56. Answer: c

The total **gross area** of a building is equal to the sum of the net area and secondary area.

An area for a specific user activity is called the **net area**, or **net assignable area**.

The remaining spaces of a building area called **unassigned area**, or **secondary space,** include circulation space, mechanical and electrical rooms, shafts, structural and wall thickness, etc.

The following are incorrect answers:
- the sum of the net area and the building wall thickness
- the sum of the net area and the assigned area
- none of the above

57. Answer: a and b
The following will affect a project's budget:
- location of the project
- inflation

The following are less likely to affect a project's budget, and are therefore the incorrect answers:
- schedule of the city's architectural review board
- traffic conditions near the site

58. Answer: b

This building will cost 4.33 million dollars in City B.

(1663/1268) x 3.3 million dollars = 4.33 million dollars

59. Answer: a, b and c

The following will affect a project's schedule:
- the client (The client's ability to make decision in a timely manner has a significant impact on the schedule.)
- the number of project team members (More people in the team can speed up the project.)
- the architect (The experience of the architect or her associates will affect a project's schedule.)

Insurance is less likely to affect a project's schedule.

60. Answer: a, b and d

The following will restrict a project's development:
- the municipal codes
- the building codes
- the fire department

The experience of construction workers is less likely to restrict a project's development.

61. Answer: a, b, c and d

The following are major considerations during programming:
- function
- form
- time
- economy

The following are less significant when compared with the four previously mentioned factors:
- neighborhood
- green building features

62. Answer: b and d

A soils report typically includes the following:
- seismic considerations
- footing and foundation design guidelines

A soils report typically does not include the following:
- landscaping
- utilities

Please note the difference between a soils report and a site survey. A **soils report** is typically prepared by a soils engineer. Its main purpose is to explore the underground soil conditions and provide design recommendations and guidelines for the structural engineer.

A **site survey** is typically prepared by a land surveyor or a civil engineer. Its main purpose is to document the existing surface conditions at the site, including contours, landscaping, easements, existing buildings and utilities, etc.

63. Answer: c

The following is the correct order for arranging the units used in the Public Land Survey System (PLSS) in the US, from large to small:

- check, township, section

A **check** is a square parcel of land 576 square miles in area with 24-mile-long sides.
A **township** is a square parcel of land 36 square miles in area with 6-mile-long sides.
A **section** is a square parcel of land 1 square mile in area with 1-mile-long sides.

64. Answer: a and b

The following are measures to improve water quality:

- **retention pond** or **retention basin** (This is also called a **wet pond, wet detention basin**, or **lake fail**, and is used to hold water permanently.) See the following link: http://en.wikipedia.org/wiki/Retention_basin
- **detention pond** or **detention basin** (This is also called a **dry pond, holding pond**, or, **dry detention basin** if no permanent pool of water exists. It is used to hold water for a limited time.) See the following link: http://en.wikipedia.org/wiki/Detention_basin

The following are incorrect answers:
- concrete swale
- trench drain

65. Answer: b

The following is the best method to reduce asbestos exposure in an existing building:
- Keep asbestos-containing materials in place without disturbing it.

The following are incorrect answers:
- Remove the asbestos. (This can disturb the asbestos-containing materials and make matter worse.)
- Seal off the spaces with asbestos containing materials. (This is not practical.)
- Ban children from rooms with asbestos containing materials. (This is not practical.)

66. Answer: a
The 90-degree parking configuration will accommodate most cars per 100 linear feet of curb. See the following details.
- 90 degree parking will accommodate about 11 cars per 100 lineal feet of curb.
- 60 degree parking will accommodate about 9 cars per 100 lineal feet of curb.
- 45 degree parking will accommodate about 8 cars per 100 lineal feet of curb.
- 30 degree parking will accommodate about 5 cars per 100 lineal feet of curb.

67. Answer: d
Pay attention to the word "not."
ADA does <u>not</u> regulate a building's height, and is therefore the correct answer.

The following are incorrect answers:
- zoning codes
- building codes
- CC&R (Covenants, Conditions and Restrictions)

68. Answer: d
Pay attention to the word "not."

The following do <u>not</u> need to comply with ADA Accessibility Guidelines (ADAAG), and are therefore the correct answers:
- a single family home on a private property
- a construction trailer

A city hall building needs to comply with ADA Accessibility Guidelines (ADAAG).

69. Answer: a and c
In a construction project, a contractor's insurance will pay for claims from the owner, if the owner waives his rights to sue for and recover from the contractor. This arrangement is an example of a **waiver of subrogation**, or an **exculpatory clause**.

The following are incorrect answers:
- an exclusive right
- a waiver of abrogation (This is a distracter, and an invented term.)

70. Answer: a and d
Per AIA Document B101-2007, the following regarding mediation is true:
- Mediation shall always precede litigation. (See AIA Document B101-2007, 8.2.1, 8.2.4)
- The parties shall share the mediation fees equally. (See AIA Document B101-2007, 8.2.3)

The following are incorrect answers:
- The prevailing party will be reimbursed for the mediation fees.
- Arbitration shall always precede mediation.

71. Answer: a
Per AIA Document B101-2007, 3.2.1, the owner should furnish the program for a project.

The following are incorrect answers:
- architect
- contractor
- building official

72. Answer: a and c
Eminent domain means the state has the power to seize a citizen's private property, without the owner's consent and use the property for the public benefits. The state has to pay a fair market value to the owner for the loss of the property.

See the following link:
http://en.wikipedia.org/wiki/Eminent_domain

The following can use eminent domain to acquire land for project development:
- a public school
- an interstate freeway

The following are incorrect answers:
- a toll road (These are typically owned by a private owner instead of a public entity.)
- a shopping center

73. Answer: b
Rehabilitation receives the most tax incentive with regard to historical buildings.

See the following related link:
http://www.nps.gov/tps/tax-incentives/taxdocs/about-tax-incentives.pdf
http://www.nps.gov/history/hps/tps/tax/rehabstandards.htm

- **Preservation** is the process of preserving monuments, buildings, etc.
- **Rehabilitation** is defined as "the process of returning a property to a state of utility, through repair or alteration, which makes possible an efficient contemporary use while preserving those portions and features of the property which are significant to its historic, architectural, and cultural values." The Standards for Rehabilitation (codified in 36 CFR 67 for use in the Federal Historic Preservation Tax Incentives program) address this most prevalent treatment.
- **Restoration** is work performed on a building to return it to a previous state.
- **Reconstruction** is the process of rebuilding a historic building that has been destroyed, based on historic records, etc.

74. Answer: a

The following is the correct order to arrange building types based on their building efficiencies, from high to low:

- department store, office, apartment, hospital

Building efficiency is the ratio of area that is assigned to a function (NASF) to gross square feet (GSF).

Building efficiency = NASF/GSF

A building with less supportive spaces is more efficient, and has a high building efficiency.

75. Answer: a

Buildings codes, ADA, municipal codes, and EPA Codes of Federal Regulations are laws, because they have gone through the legislation process, but reference guides by USGBC are NOT laws. They are rules set by the USGBC. The USGBC has NO legal authority like the other governing agencies.

LEED standards are voluntary. You choose to obey the rules when you seek certification for a building, but these rules are NOT laws.

Note:

NCARB has started to draw 23% to 29% or 19 to 25 of the questions of the PPP ARE exam from Environmental, Social & Economic Issues. We include some questions on green buildings and LEED to help you to meet this new challenge.

If you are weak on this subject, you are more than welcome to check out our LEED Exam Guide series and other books. They are available as printed books and PDF eBooks at our website:
http://www.GreenExamEducation.com

You can preview up to 20% of all our books' content at:
http://books.google.com/

Please also see our website for more FREE tips and downloads:
http://www.GreenExamEducation.com

76. Answer: c, d, and e

Increasing the site coverage will increase impervious area and will increase stormwater runoff. Increasing the FAR may or may NOT increase impervious area. Porous pavement will help recharge the groundwater thereby reducing stormwater runoff, and high-albedo (high-reflectivity) materials will increase reflectivity to alleviate the urban "heat island" effect. Vegetated roofs and retention ponds can also reduce stormwater runoff and alleviate the urban "heat island" effect.

77. Answer: b
Recycled materials can protect virgin materials, but may require more energy to process, can increase traffic, and increase MEP cost.

78. Answer: f
Graywater is the household water that has not come into contact with the kitchen sink or toilet waste.

See USGBC Definitions at the link below:

https://www.usgbc.org/ShowFile.aspx?DocumentID=5744

79. Answer: a and d
Read the question carefully; it is asking for the WRONG statements.

80. Answer: a and d
Open spaces must be pervious and vegetated. They are non-built environments.

81. Answer: e
Life cycle cost analysis is unique to a LEED certified building. Hard costs, soft costs, and storm control costs are required by both a LEED certified building and a conventional building. "Life cycle analysis costs" is different from life cycle cost analysis, and is a distracter.

Life cycle cost analysis is an evaluation of a building's economic performance including operational and maintenance costs over the life of the product.

Life cycle analysis is the same as eco-balance, cradle-to-cradle analysis, or life cycle assessment. It is used to evaluate the environmental impact of a service or product.

82. Answer: d
Minimum energy performance is a prerequisite, not a credit. Emission measurement and radon alleviation are distracters.

83. Answer: c
Concerning ozone depletion potential (ODP): CFCs>HCFCs>HFCs. Therefore, HFCs have the lowest ODP.

84. Answer: a
Construction debris sent to a recycle facility is the only item that has been used by a consumer, and is therefore a post-consumer item.

85. Answer: b
ISO 14000 evaluates the environmental performance of services and products. It includes Design for Environment, Life Cycle Assessment, and Environmental Labels and Declaration.

B. Mock Exam Solution: Site Zoning Vignette

1. A step-by-step solution to the graphic vignette: Site Zoning

1) Surface improvements are prohibited within 6 ft of any property line. Use **Sketch > Rectangle** to draw a number of rectangles defining the no improvement areas within 6 ft of all property lines (figure 4.1).

2) Construction of buildings and other surface improvements is prohibited within 20 ft of the lake high water line. Use **Sketch > Circle** to draw a number of circles with 20 ft radii defining the no improvement areas within 20 ft of the lake high water line. The centers of the circles should be placed on the lake high water line (figure 4.2).

   Note: The radius of the circle displays on the lower-left-hand corner of the screen when drawn. After you draw the first circle, the radius for rest of the circles will stay at 20 ft, and you just need to click on the lake high water line to place them.

3) Use **Draw > Secondary Construction Area** to draw the surface improvement areas (figure 4.3).

   Note: The secondary construction area is a polygon. You draw a starting point, click for each corner of the polygon, and then close the polygon by clicking on the starting point again.

4) Use **Sketch > Rectangle** to draw a number of rectangles defining the building setbacks from all property lines (figure 4.4). Front yard setbacks are only considered from Main Street.
   Front yard setbacks from property line along Main Street: 20 ft
   Rear yard setbacks: 25 ft
   Side yard setbacks: 10 ft

5) Construction of buildings is prohibited within the existing drainage easement. Use **Draw > Buildable Area** to draw the buildable areas (figure 4.5).

   Note: The building construction area is also a polygon and drawn as described in step three.

   In this case, the existing drainage easement is MORE restrictive than the 10 ft side yard setback requirement, so we comply with the more restrictive requirement.

6) Draw the profile of the existing grade at Section A-A:
   - Use **Sketch > Line** to project the intersections of the contour lines and section line of A-A to the view below (figure 4.6).
   - Per the elevations of the contour lines in plan, use **Draw > Grade** to draw the profile of the existing grade at Section A-A on the grid (figure 4.7).

Note: *The key of solving this vignette is locating the benchmark and its elevation. The benchmark is located at the intersection of the lake high water line and the west property line of Lot B. Its elevation is 110', the same as the lake high water line.*

7) The maximum building height limit within 55 ft of the east property line of Lot B shall be 50 ft above the benchmark elevation. Since the benchmark elevation is 110', we can calculate this as: 110' + 50' = 160'. Use **Sketch > Rectangle** to draw a rectangle showing this criterion (figure 4.8).

8) The maximum building height limit between 0 ft and 40 ft of the west property line in Lot A shall be 25 ft above the grade at the property line. The grade at the intersection of the east property line in Lot A and the section line of Section A-A is 150'. We can calculate this as: 150' + 25' = 175'. Use **Sketch > Rectangle** to draw a rectangle showing this requirement (figure 4.9).

9) The maximum building height limit shall be 80 ft above the benchmark elevation. Since the benchmark elevation is 110', we can calculate this as: 110' + 80' = 190'. Use **Sketch > Rectangle** to draw a rectangle showing this criterion (figure 4.10).

10) The maximum building envelope is restricted to an elevation defined by a 45-degree line rising westward from a point at an elevation of 25 ft directly above the benchmark. Since the benchmark elevation is 110', we can calculate this as: 110' + 25' = 135'. Use **Sketch > Line** to draw a line showing this criterion (figure 4.11).

Note: *The angle of the line displays on the lower-left-hand corner as you draw. Getting the angle exactly at 135 degrees is almost impossible, because of NCARB software limitation. The angle should be drawn larger than 135 degrees rather than smaller than 135 degrees, because it is more restrictive and on the safe side.*

11) Use **Sketch > Line** to draw lines projecting down from the buildable areas *at section line A-A* in the floor plan (figure 4.12).

Note: *The program explicitly states: "On the grid, draw the profile of the maximum building envelope for each lot at Section A-A."*

It is VERY easy to make the mistake of drawing the building profile all the way to the side yard setback lines, especially after program requires you to draw the 135-degree line from the property line.

12) Use **Draw > Building Profile** to draw the building profiles which comply with all the criteria listed in previous steps 7 to 11 (figure 4.13).

13) Use **Zoom** to zoom out. Use **Sketch > Hide Sketch Elements** to hide sketch elements. This is the final solution (figure 4.14).

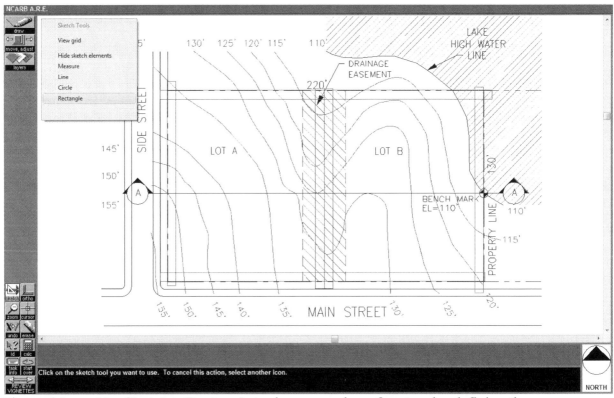

Figure 4.1   Use **Sketch** > **Rectangle** to draw a number of rectangles defining the no improvement areas within 6 ft of all property lines.

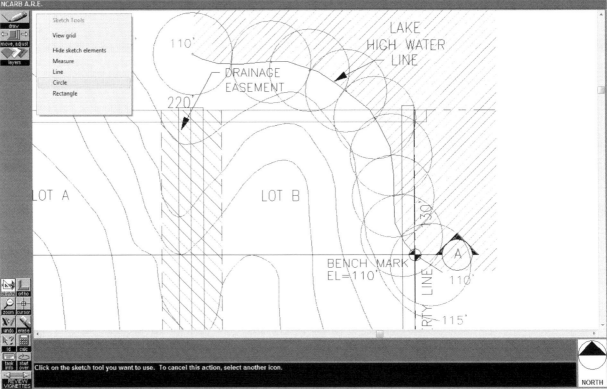

Figure 4.2   Use **Sketch** > **Circle** to draw a number of circles with 20 ft radii defining the no improvement areas within 20 ft of the lake high water line.

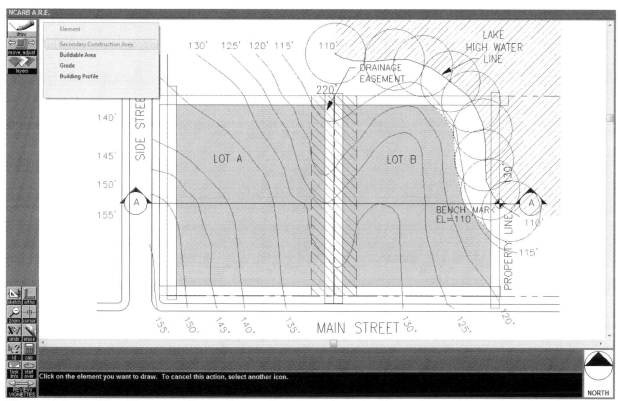

Figure 4.3   Use **Draw > Secondary Construction Area** to draw the surface improvement
areas.

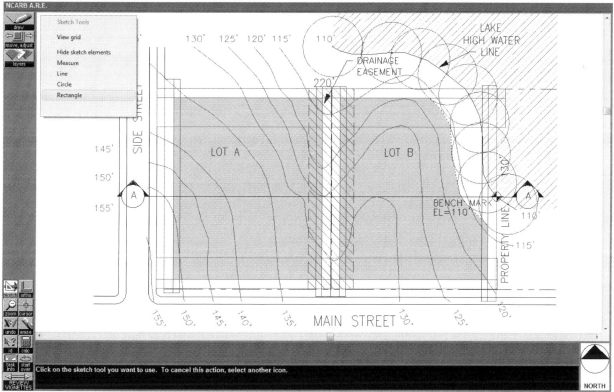

Figure 4.4   Use **Sketch > Rectangle** to draw a number of rectangles defining the building
setbacks from all property lines.

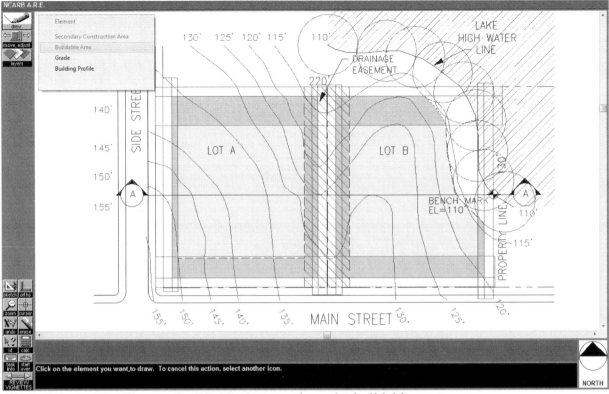

Figure 4.5   Use **Draw > Buildable Area** to draw the buildable areas.

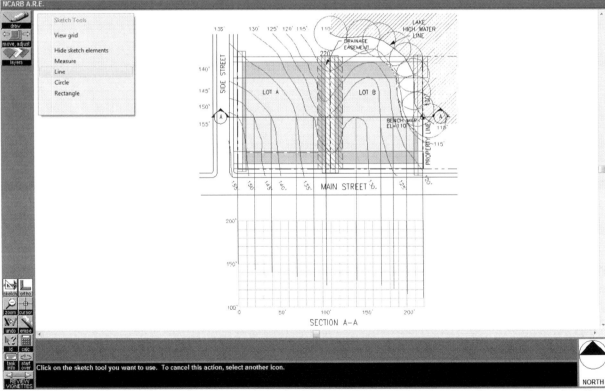

Figure 4.6   Use **Sketch > Line** to project the intersections of the contour lines and section line A-A to the grid below.

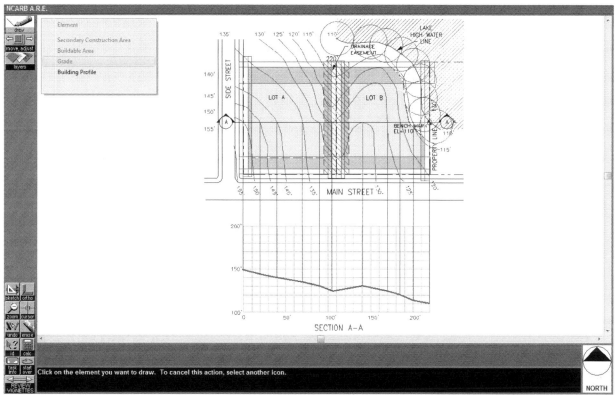

Figure 4.7   Per the elevations of the contour lines in plan, use **Draw > Grade** to draw the profile of the existing grade at Section A-A on the grid.

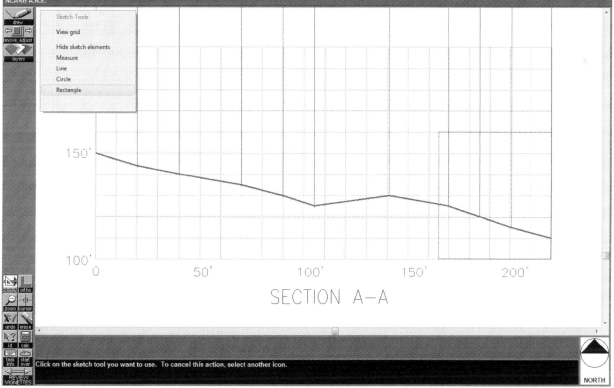

Figure 4.8   Use **Sketch > Rectangle** to draw a rectangle to showing the maximum building height limit within 55 ft of the east property line of Lot B.

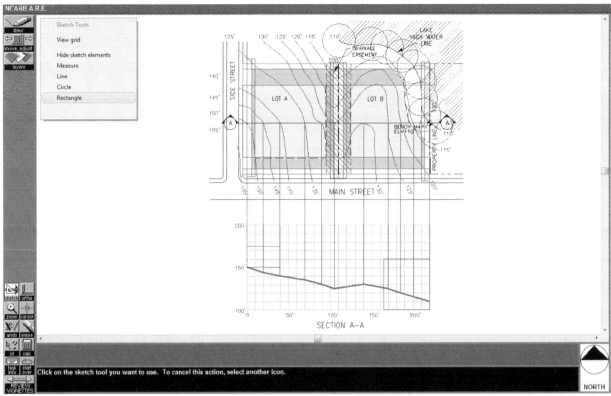

Figure 4.9   Use **Sketch > Rectangle** to draw a rectangle showing the maximum building height limit between 0 ft and 40 ft of the west property line in Lot A.

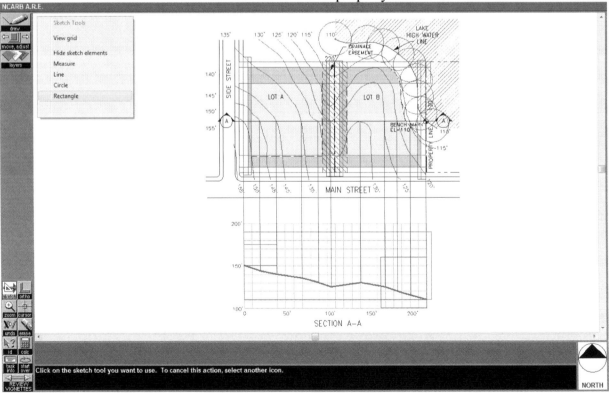

Figure 4.10   Use **Sketch > Rectangle** to draw a rectangle showing that the maximum building height limit shall be 80 ft above the benchmark elevation.

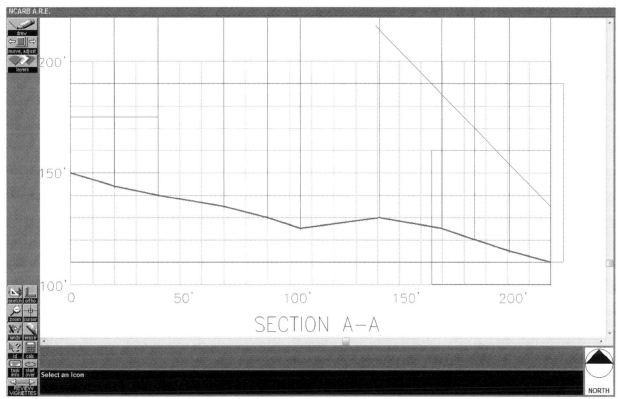

Figure 4.11   Use **Sketch > Line** to draw a line showing that the maximum building envelope is restricted to an elevation defined by a 135-degree line.

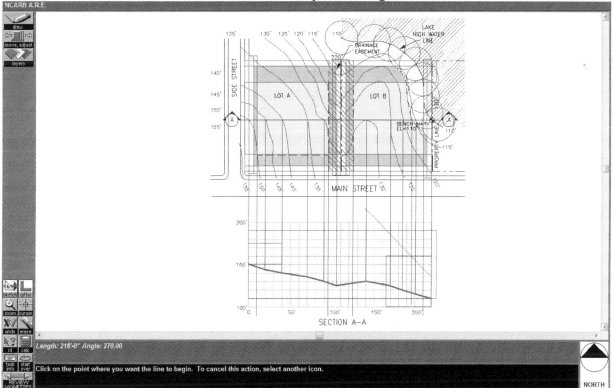

Figure 4.12   Use **Sketch > Line** to draw lines projecting down from the buildable areas at section line A-A in the floor plan.

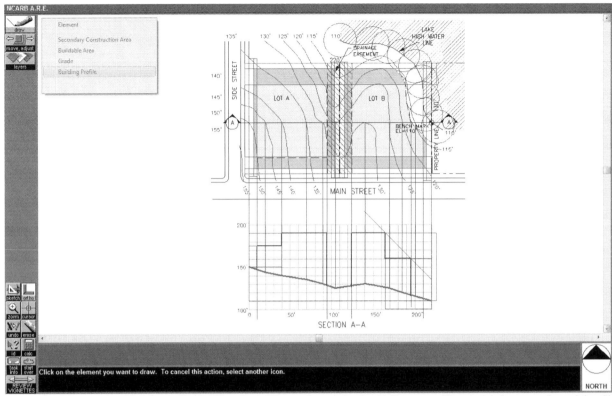

Figure 4.13   Use **Draw > Building Profile** to draw the building profiles which comply with all the criteria listed in previous steps.

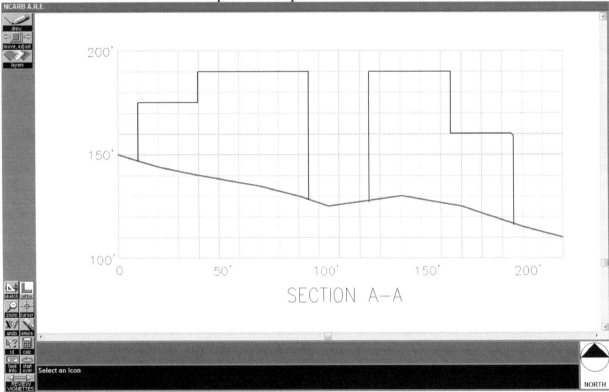

Figure 4.14   Use **Sketch > Hide Sketch Elements** to hide sketch elements. This is the final solution.

2. Notes on mock exam graphic vignette traps
Several **common errors** or **traps** which you may fall into are:

1) The **existing drainage easement** is MORE restrictive than the 10-ft side yard setback requirement, so you need to comply with the more restrictive requirement for the buildable areas and building profiles.

2) The existing drainage easement area can still be a **secondary construction area**.

3) The key of solving this vignette is locating the **benchmark** and its elevation. The benchmark is located at the intersection of the lake high water line and the East property line of Lot B. Its elevation is 110', the same as the lake high water line.

4) The program states, "On the grid, draw the profile of the maximum building envelope for each lot **at Section A-A**." It is VERY easy to make the mistake of drawing the building profile all the way to the side yard setback lines, especially after the program requires you to draw the 135-degree height restriction angle from a property line.

Appendixes

A. List of Figures

B. Official Reference Materials Suggested by NCARB

1. General NCARB reference materials for ARE:

Per NCARB, all candidates should become familiar with the latest version of the following codes:

International Code Council, Inc. (ICC, 2006)
International Building Code
International Mechanical Code
International Plumbing Code

National Fire Protection Association (NFPA)
Life Safety Code (NFPA 101)
National Electrical Code (NFPA 70)

National Research Council of Canada
National Building Code of Canada
National Plumbing Code of Canada
National Fire Code of Canada

American Institute of Architects
AIA Documents - 2007

Candidates should be familiar with the Standard on Accessible and Usable Buildings and Facilities (ICC/ANSI A117.1-98)

2. Official NCARB reference materials for the Programming, Planning & Practice (PPP) division:

The Architect's Handbook of Professional Practice
Joseph A. Demkin, AIA, Executive Editor
The American Institute of Architects
John Wiley & Sons, latest edition
This comprehensive book covers all aspect of architectural practice, and includes two CDs containing the sample AIA contract documents.

Architectural Graphic Standards
Charles G. Ramsey and Harold R. Sleeper
The American Institute of Architects
John Wiley & Sons, latest edition

Canadian Handbook of Practice for Architects,
Committee of Canadian Architectural Councils and The Royal Architectural Institute of Canada, latest edition

Design With Climate
Victor Olgyay
Van Nostrand Reinhold, 1992

Design With Nature
Ian L. McHarg
John Wiley & Sons, 1992

Designing Places for People
C. M. Deasy, FAIA
Whitney Library of Design, 1990

A History of Architecture: Settings & Rituals
Spiro Kostoff
Oxford University Press, 1995

The Image of the City
Kevin Lynch
MIT Press, 1960

Modern Architecture
Alan Colquhoun
Oxford University Press, 2002

The New Urbanism
Peter Katz
McGraw-Hill, 1994

A Pattern Language: Towns, Buildings, Construction
Christopher Alexander, Sarah Ishikawa, and Murray Silverstein
Oxford University Press, 1977

Programming for Design: From Theory to Practice
Edith Cherry
John Wiley & Sons, 1998

Sir Banister Fletcher's A History of Architecture
John Musgrove, Editor
Butterworths-Heinmann, 1996

Site Planning, Third Edition
Kevin Lynch and Gary Hack
MIT Press, 1984

*Suburban Nation: The Rise of Sprawl and the
Decline of the American Dream*
Andres Duany, Elizabeth Plater-Zybeck, and Jeff Speck
North Point Press, 2001

Sustainable Design Fundamentals for Buildings
National Practice Program
Canada, 2001

C. Other Reference Materials

Chen, Gang. *LEED GA Exam Guide: A Must-Have for the LEED Green Associate Exam: Comprehensive Study Materials, Sample Questions, Mock Exam, Green Building LEED Certification, and Sustainability.* ArchiteG, Inc., latest edition.
This book is a good introduction to green buildings and the LEED building rating systems.

Ching, Francis. *Architecture: Form, Space, & Order.* Wiley, latest edition.
This is one of the best architectural books that you can have. I still flip through it every now and then. It is a great source for inspiration.

Frampton, Kenneth. *Modern Architecture: A Critical History.* Thames and Hudson, London, latest edition.
The publication is a valuable resource for architectural history.

Jarzombek, Mark M. (Author), Vikramaditya Prakash (Author), Francis D. K. Ching (Editor). *A Global History of Architecture.* Wiley, latest edition.
Filled with 1000 b & w photos, 50 color photos, and 1500 b & w illustrations, this is a valuable and comprehensive resource for architectural history. It does not limit the topic to a Western perspective, but rather gives a global perspective.

Trachtenberg, Marvin and Isabelle Hyman. *Architecture: From Pre-history to Post-Modernism.* Prentice Hall, Englewood Cliffs, NJ latest edition.
This is also a valuable and comprehensive resource for architectural history.

D. Definition of "Architect" and Some Important Information about Architects and the Profession of Architecture

Architects, Except Landscape and Naval
- Nature of the Work
- Training, Other Qualifications, and Advancement
- Employment
- Job Outlook
- Projections Data
- Earnings
- OES Data
- Related Occupations
- Sources of Additional Information

Significant Points
- About one in five architects are self-employed—more than two times the proportion for all occupations.
- Licensing requirements include a professional degree in architecture, at least three years of practical work training, and passing all divisions of the Architect Registration Examination.
- Architecture graduates may face competition, especially for jobs in the most prestigious firms.

Nature of the Work
People need places in which to live, work, play, learn, worship, meet, govern, shop, and eat. These places may be private or public; indoors or out; rooms, buildings, or complexes, and architects are the individuals who design them. Architects are licensed professionals trained in the art and science of building design who develop the concepts for structures and turn those concepts into images and plans.

Architects create the overall aesthetic and look of buildings and other structures, but the design of a building involves far more than its appearance. Buildings must also be functional, safe, economical, and must suit the needs of the people who use them. Architects consider all these factors when they design buildings and other structures.
Architects may be involved in all phases of a construction project, from the initial discussion with the client through the entire construction process. Their duties require specific skills—designing, engineering, managing, supervising, and communicating with clients and builders. Architects spend a great deal of time explaining their ideas to clients, construction contractors, and others. Successful architects must be able to communicate their unique vision persuasively.

The architect and client discuss the objectives, requirements, and budget of a project. In some cases, architects provide various pre-design services: conducting feasibility and environmental impact studies, selecting a site, preparing cost analysis and land-use studies, or specifying the requirements the design must meet. For example, they may determine space requirements by researching the numbers and types of potential users of a building. The architect then prepares drawings and a report presenting ideas for the client to review.

After discussing and agreeing on the initial proposal, architects develop final construction plans that show the building's appearance and details for its construction. Accompanying these plans are drawings of the structural system; air-conditioning, heating, and ventilating systems; electrical systems; communications systems; plumbing; and, possibly, site and landscape plans. The plans also specify the building materials and, in some cases, the interior furnishings. In developing designs, architects follow building codes, zoning laws, fire regulations, and other ordinances, such as those requiring easy access by people who are disabled. Computer-aided design and drafting (CADD) and Building Information Modeling (BIM) technology has replaced traditional paper and pencil as the most common method for creating design and construction drawings. Continual revision of plans on the basis of client needs and budget constraints is often necessary.

Architects may also assist clients in obtaining construction bids, selecting contractors, and negotiating construction contracts. As construction proceeds, they may visit building sites to make sure that contractors follow the design, adhere to the schedule, use the specified materials, and meet work quality standards. The job is not complete until all construction is finished, required tests are conducted, and construction costs are paid. Sometimes, architects also provide post-construction services, such as facilities management. They advise on energy efficiency measures, evaluate how well the building design adapts to the needs of occupants, and make necessary improvements.

Often working with engineers, urban planners, interior designers, landscape architects, and other professionals, architects in fact spend a great deal of their time coordinating information from, and the work of, other professionals engaged in the same project.

They design a wide variety of buildings, such as office and apartment buildings, schools, churches, factories, hospitals, houses, and airport terminals. They also design complexes such as urban centers, college campuses, industrial parks, and entire communities.

Architects sometimes specialize in one phase of work. Some specialize in the design of one type of building—for example, hospitals, schools, or housing. Others focus on planning and pre-design services or construction management and do minimal design work.

Work environment. Usually working in a comfortable environment, architects spend most of their time in offices consulting with clients, developing reports and drawings, and working with other architects and engineers. However, they often visit construction sites to review the progress of projects. Although most architects work approximately 40 hours per week, they often have to work nights and weekends to meet deadlines.

Training, Other Qualifications, and Advancement
There are three main steps in becoming an architect. First is the attainment of a professional degree in architecture. Second is work experience through an internship, and third is licensure through the passing of the Architect Registration Exam.

Education and training. In most states, the professional degree in architecture must be from one of the 114 schools of architecture that have degree programs accredited by the National Architectural Accrediting Board. However, state architectural registration boards set their own standards, so graduation from a non-accredited program may meet the educational requirement for licensing in a few states.

Three types of professional degrees in architecture are available: a five year bachelor's degree, which is most common and is intended for students with no previous architectural training; a two year master's degree for students with an undergraduate degree in architecture or a related area; and a three or four year master's degree for students with a degree in another discipline.

The choice of degree depends on preference and educational background. Prospective architecture students should consider the options before committing to a program. For example, although the five year bachelor of architecture offers the fastest route to the professional degree, courses are specialized, and if the student does not complete the program, transferring to a program in another discipline may be difficult. A typical program includes courses in architectural history and theory, building design with an emphasis on CADD, structures, technology, construction methods, professional practice, math, physical sciences, and liberal arts. Central to most architectural programs is the design studio, where students apply the skills and concepts learned in the classroom, creating drawings and three-dimensional models of their designs.

Many schools of architecture also offer post-professional degrees for those who already have a bachelor's or master's degree in architecture or other areas. Although graduate education beyond the professional degree is not required for practicing architects, it may be required for research, teaching, and certain specialties.

All state architectural registration boards require architecture graduates to complete a training period—usually at least three years—before they may sit for the licensing exam. Every state, with the exception of Arizona, has adopted the training standards established by the Intern Development Program, a branch of the American Institute of Architects and the National Council of Architectural Registration Boards (NCARB). These standards stipulate broad training under the supervision of a licensed architect. Most new graduates complete their training period by working as interns in architectural firms. Some States allow a portion of the training to occur in the offices of related professionals, such as engineers or general contractors. Architecture students who complete internships while still in school can count some of that time toward the three year training period.

Interns in architectural firms may assist in the design of one part of a project, help prepare architectural documents or drawings, build models, or prepare construction drawings on CADD. Interns also may research building codes and materials or write specifications for building materials, installation criteria, the quality of finishes, and other related details.

Licensure. All states and the District of Columbia require individuals to be licensed (registered) before they may call themselves architects and contract to provide architectural services. During the time between graduation and becoming licensed, architecture school graduates generally work in the field under the supervision of a licensed architect who takes legal responsibility for all work. Licensing requirements include a professional degree in architecture, a period of practical training or internship, and a passing score on all divisions of the Architect Registration Examination. The examination is broken into seven divisions consisting of either multiple choice and/or graphic vignettes. The eligibility period for completion of all divisions is five years from the date of passing your first exam.

Most states also require some form of continuing education to maintain a license, and many others are expected to adopt mandatory continuing education. Requirements vary by state but usually involve the completion of a certain number of credits annually or biennially through workshops, formal university classes, conferences, self-study courses, or other sources.

Other qualifications. Architects must be able to communicate their ideas visually to their clients. Artistic and drawing ability is helpful, but not essential, to such communication. More important are a visual orientation and the ability to understand spatial relationships. Other important qualities for anyone interested in becoming an architect are creativity and the ability to work independently and as part of a team. Computer skills are also required for writing specifications, for two and three dimensional drafting using CADD programs, and for financial management.

Certification and advancement. A growing number of architects voluntarily seek certification by the National Council of Architectural Registration Boards. Certification is awarded after independent verification of the candidate's educational transcripts, employment record, and professional references. Certification can make it easier to become licensed across states. In fact, it is the primary requirement for reciprocity of licensing among state boards that are NCARB members. In 2007, approximately one-third of all licensed architects had this certification.

After becoming licensed and gaining experience, architects take on increasingly responsible duties, eventually managing entire projects. In large firms, architects may advance to supervisory or managerial positions. Some architects become partners in established firms, while others set up their own practices. Some graduates with degrees in architecture also enter related fields, such as graphic, interior, or industrial design; urban planning; real estate development; civil engineering; and construction management.

Employment
Architects held about 132,000 jobs in 2006. Approximately seven out of ten jobs were in the architectural, engineering, and related services industry—mostly in architectural firms with fewer than five workers. A small number worked for residential and nonresidential building construction firms and for government agencies responsible for housing, community planning, or construction of government buildings, such as the U.S. Departments of Defense and Interior, and the General Services Administration. About one in five architects are self-employed.

Job Outlook
Employment of architects is expected to grow faster than the average for all occupations through 2016. Keen competition is expected for positions at the most prestigious firms, and

opportunities will be best for those architects who are able to distinguish themselves with their creativity.

Employment change. Employment of architects is expected to grow by 18 percent between 2006 and 2016, which is <u>faster than the average</u> for all occupations. Employment of architects is strongly tied to the activity of the construction industry. Strong growth is expected to come from nonresidential construction as demand for commercial space increases. Residential construction, buoyed by low interest rates, is also expected to grow as more people become homeowners. If interest rates rise significantly, home building may fall off, but residential construction makes up only a small part of architects' work.

Current demographic trends also support an increase in demand for architects. As the population of Sunbelt States continues to grow, the people living there will need new places to live and work. As the population continues to live longer and baby-boomers begin to retire, there will be a need for more healthcare facilities, nursing homes, and retirement communities. In education, buildings at all levels are getting older and class sizes are getting larger. This will require many school districts and universities to build new facilities and renovate existing ones.

In recent years, some architecture firms have outsourced the drafting of construction documents and basic design for large-scale commercial and residential projects to architecture firms overseas. This trend is expected to continue and may have a negative impact on employment growth for lower level architects and interns who would normally gain experience by producing these drawings.

Job prospects. Besides employment growth, additional job openings will arise from the need to replace the many architects who are nearing retirement, and others who transfer to other occupations or stop working for other reasons. Internship opportunities for new architectural students are expected to be good over the next decade, but more students are graduating with architectural degrees and some competition for entry-level jobs can be anticipated. Competition will be especially keen for jobs at the most prestigious architectural firms as prospective architects try to build their reputation. Prospective architects who have had internships while in school will have an advantage in obtaining intern positions after graduation. Opportunities will be best for those architects that are able to distinguish themselves from others with their creativity.

Prospects will also be favorable for architects with knowledge of "green" design. Green design, also known as sustainable design, emphasizes energy efficiency, renewable resources such as energy and water, waste reduction, and environmentally friendly design, specifications, and materials. Rising energy costs and increased concern about the environment has led to many new buildings being built green.

Some types of construction are sensitive to cyclical changes in the economy. Architects seeking design projects for office and retail construction will face especially strong competition for jobs or clients during recessions, and layoffs may ensue in less successful firms. Those involved in the design of institutional buildings, such as schools, hospitals, nursing homes, and correctional facilities, will be less affected by fluctuations in the economy. Residential construction makes up a small portion of work for architects, so major changes in the housing market would not be as significant as fluctuations in the nonresidential market.

Despite good overall job opportunities, some architects may not fare as well as others. The profession is geographically sensitive, and some parts of the Nation may have fewer new building projects. Also, many firms specialize in specific buildings, such as hospitals or office towers, and demand for these buildings may vary by region. Architects may find it increasingly necessary to gain reciprocity in order to compete for the best jobs and projects in other states.

Projections Data

Projections data from the National Employment Matrix

| Occupational title | SOC Code | Employment, 2006 | Projected employment, 2016 | Change, 2006-16 | | Detailed statistics |
				Number	Percent	
Architects, except landscape and naval	17-1011	132,000	155,000	23,000	18	PDF zipped XLS

NOTE: Data in this table are rounded. See the discussion of the employment projections table in the *Handbook* introductory chapter on *Occupational Information Included in the Handbook*.

Earnings
Median annual earnings of wage-and-salary architects were $64,150 in May 2006. The middle 50 percent earned between $49,780 and $83,450. The lowest 10 percent earned less than $39,420, and the highest 10 percent earned more than $104,970. Those just starting their internships can expect to earn considerably less.

Earnings of partners in established architectural firms may fluctuate because of changing business conditions. Some architects may have difficulty establishing their own practices and may go through a period when their expenses are greater than their income, requiring substantial financial resources.

Many firms pay tuition and fees toward continuing education requirements for their employees.

For the latest wage information:
The above wage data is from the Occupational Employment Statistics (OES) survey program, unless otherwise noted. For the latest national, state, and local earnings data, visit the following pages:

Architects, except landscape and naval

Related Occupations

Architects design buildings and related structures. Construction managers, like architects, also plan and coordinate activities concerned with the construction and maintenance of buildings and facilities. Others who engage in similar work are landscape architects, civil engineers, urban and regional planners, and designers, including interior designers, commercial and industrial designers, and graphic designers.

Sources of Additional Information

Disclaimer:
Links to non-BLS Internet sites are provided for your convenience and do not constitute an endorsement.

Information about education and careers in architecture can be obtained from:
- The American Institute of Architects, 1735 New York Ave. NW., Washington, DC 20006. Internet: http://www.aia.org
- Intern Development Program, National Council of Architectural Registration Boards, Suite 1100K, 1801 K St. NW., Washington, D.C. 20006. Internet: http://www.ncarb.org OOH ONET Codes 17-1011.00"

Quoted from: Bureau of Labor Statistics, U.S. Department of Labor, Occupational Outlook Handbook, 2008-09 Edition, Architects, Except Landscape and Naval, on the Internet at **http://www.bls.gov/oco/ocos038.htm** (visited November 30, 2008).
Last Modified Date: December 18, 2007

Note: Please check the website above for the latest information.

E. AIA Compensation Survey

Every three years, AIA publishes a Compensation Survey for various positions at architectural firms across the country. It is a good idea to find out the salary before you make the final decision to become an architect. If you are already an architect, it is also a good idea to determine if you are underpaid or overpaid.

See following link for some sample pages for the 2008 AIA Compensation Survey:

http://www.aia.org/aiaucmp/groups/ek_public/documents/pdf/aiap072881.pdf

F.   So ... You would Like to Study Architecture

To study architecture, you need to learn how to draft, how to understand and organize spaces and the interactions between interior and exterior spaces, how to do design, and how to communicate effectively. You also need to understand the history of architecture.

As an architect, a leader for a team of various design professionals, you not only need to know architecture, but also need to understand enough of your consultants' work to be able to coordinate with them. Your consultants include soils and civil engineers, landscape architects, structural, electrical, mechanical, and plumbing engineers, interior designers, sign consultants, etc.

There are two major career paths for you in architecture: practice as an architect or teach in college or university. The earlier you determine which path you are going to take, the more likely you will be successful at an early age. Some famous and well-respected architects, like my USC alumnus Frank Gehry, have combined the two paths successfully. They teach at the universities and have their own architectural practice. Even as a college or university professor, people respect you more if you have actual working experience and have some built projects. If you only teach in colleges or universities but have no actual working experience and have no built projects, people will consider you as a "paper" architect, and they are not likely to take you seriously, because they will think you probably do not know how to put a real building together.

In the U.S., if you want to practice architecture, you need to obtain an architect's license. It requires a combination of passing scores on the Architectural Registration Exam (ARE) and eight years of education and/or qualified working experience, including at least one year of working experience in the U.S. Your working experience needs to be under the supervision of a licensed architect to be counted as qualified working experience for your architect's license.

If you work for a landscape architect or civil engineer or structural engineer, some states' architectural licensing boards will count your experience at a discounted rate for the qualification of your architect's license. For example, two years of experience working for a civil engineer may be counted as one year of qualified experience for your architect's license. You need to contact your state's architectural licensing board for specific licensing requirements for your state.

If you want to teach in colleges or universities, you probably want to obtain a master's degree or a Ph.D. It is not very common for people in the architectural field to have a Ph.D. One reason is that there are few Ph.D. programs for architecture. Another reason is that architecture is considered a profession and requires a license. Many people think an architect's license is more important than a Ph.D. degree. In many states, you need to have an architect's license to even use the title "architect," or the terms "architectural" or "architecture" to advertise your service. You cannot call yourself an architect if you do not have an architect's license, even if you have a Ph.D. in architecture. Violation of these rules brings punishment.

To become a tenured professor, you need to have a certain number of publications and pass the evaluation for the tenure position. Publications are very important for tenure track positions. Some people say it is "publish or perish" for the tenured track positions in universities and colleges.

The American Institute of Architects (AIA) is the national organization for the architectural profession. Membership is voluntary. There are different levels of AIA membership. Only licensed architects can be (full) AIA members. If you are an architectural student or an intern but not a licensed architect yet, you can join as an associate AIA member. Contact AIA for detailed information.

The National Council of Architectural Registration Boards (NCARB) is a nonprofit federation of architectural licensing boards. It has some very useful programs, such as IDP, to assist you in obtaining your architect's license. Contact NCARB for detailed information.

Back Page Promotion

You may be interested in some other books written by Gang Chen:

A. **ARE Mock Exam series. See the following link:**
   http://www.GreenExamEducation.com

B. **LEED Exam Guide series. See the following link:**
   http://www.GreenExamEducation.com

C. ***Building Construction:*** *Project Management, Construction Administration, Drawings, Specs, Detailing Tips, Schedules, Checklists, and Secrets Others Don't Tell You (Architectural Practice Simplified, 2nd edition)*
   http://www.ArchiteG.com

D. ***Planting Design Illustrated***
   http://outskirtspress.com/agent.php?key=11011&page=GangChen

ARE Mock Exam Series

Published ARE books:

Construction Documents and Service (CDS) Are Mock Exam (Architect Registration Exam): ARE Overview, Exam Prep Tips, Multiple-Choice Questions and Graphic Vignettes, Solutions and Explanations
ISBN-13: 9781612650005
(Published May 22, 2011)

Building Design and Construction Systems (BDCS) ARE Mock Exam (Architect Registration Exam): ARE Overview, Exam Prep Tips, Multiple-Choice Questions and Graphic Vignettes, Solutions and Explanations
ISBN-13: 9781612650029
(Published July 12, 2011)

Building Systems (BS) ARE Mock Exam (Architect Registration Exam): ARE Overview, Exam Prep Tips, Multiple-Choice Questions and Graphic Vignettes, Solutions and Explanations
ISBN-13: 9781612650036
(Published October 28, 2011)

Schematic Design (SD) ARE Mock Exam (Architect Registration Exam): ARE Overview, Exam Prep Tips, Graphic Vignettes, Solutions and Explanations
ISBN: 9781612650050
(Published November 18, 2011)

Programming, Planning & Practice (PPP) ARE Mock Exam (Architect Registration Exam): ARE Overview, Exam Prep Tips, Multiple-Choice Questions and Graphic Vignettes, Solutions and Explanations
ISBN-13: 9781612650067
(Published May 16, 2012)

Upcoming ARE books:

Other books in the ARE Mock Exam Series are being produced. Our goal is to produce one mock exam book PLUS one guidebook for each of the ARE exam division.

See the following link for the latest information:
http://www.GreenExamEducation.com

LEED Exam Guide series*:* Comprehensive Study Materials, Sample Questions, Mock Exam, Building LEED Certification and Going Green

LEED (Leadership in Energy and Environmental Design) is the most important trend of development, and it is revolutionizing the construction industry. It has gained tremendous momentum and has a profound impact on our environment.

From LEED Exam Guides series, you will learn how to

1. Pass the LEED Green Associate Exam and various LEED AP + exams (each book will help you with a specific LEED exam).

2. Register and certify a building for LEED certification.

3. Understand the intent for each LEED prerequisite and credit.

4. Calculate points for a LEED credit.

5. Identify the responsible party for each prerequisite and credit.

6. Earn extra credit (exemplary performance) for LEED.

7. Implement the local codes and building standards for prerequisites and credit.

8. Receive points for categories not yet clearly defined by USGBC.

There is currently NO official book on the LEED Green Associate Exam, and most of the existing books on LEED and LEED AP are too expensive and too complicated to be practical and helpful. The pocket guides in LEED Exam Guides series fill in the blanks, demystify LEED, and uncover the tips, codes, and jargon for LEED as well as the true meaning of "going green." They will set up a solid foundation and fundamental framework of LEED for you. Each book in the LEED Exam Guides series covers every aspect of one or more specific LEED rating system(s) in plain and concise language and makes this information understandable to all people.

These pocket guides are small and easy to carry around. You can read them whenever you have a few extra minutes. They are indispensable books for all people—administrators; developers; contractors; architects; landscape architects; civil, mechanical, electrical, and plumbing engineers; interns; drafters; designers; and other design professionals.

Why is the LEED Exam Guides series needed?

A number of books are available that you can use to prepare for the LEED Exams:

1. *USGBC Reference Guides.* You need to select the correct version of the *Reference Guide* for your exam.

   The *USGBC Reference Guides* are comprehensive, but they give too much information. For example, *The LEED 2009 Reference Guide for Green Building Design and Construction (BD&C)* has about 700 oversized pages. Many of the calculations in the books are too detailed for the exam. They are also expensive (approximately $200 each, so most people may not buy them for their personal use, but instead, will seek to share an office copy).

   It is good to read a reference guide from cover to cover if you have the time. The problem is not too many people have time to read the whole reference guide. Even if you do read the whole guide, you may not remember the important issues to pass the LEED exam. You need to reread the material several times before you can remember much of it.

   Reading the reference guide from cover to cover without a guidebook is a difficult and inefficient way of preparing for the LEED AP Exam, because you do NOT know what USGBC and GBCI are looking for in the exam.

2. The USGBC workshops and related handouts are concise, but they do not cover extra credits (exemplary performance). The workshops are expensive, costing approximately $450 each.

3. Various books published by a third party are available on Amazon, bn.com and books.google.com. However, most of them are not very helpful.

   There are many books on LEED, but not all are useful.

   LEED Exam Guides series will fill in the blanks and become a valuable, reliable source:

   a. They will give you more information for your money. Each of the books in the LEED Exam Guides series has more information than the related USGBC workshops.

   b. They are exam-oriented and more effective than the USGBC reference guides.

   c. They are better than most, if not all, of the other third-party books. They give you comprehensive study materials, sample questions and answers, mock exams and answers, and critical information on building LEED certification and going green. Other third-party books only give you a fraction of the information.

   d. They are comprehensive yet concise. They are small and easy to carry around. You can read them whenever you have a few extra minutes.

   e. They are great timesavers. I have highlighted the important information that you need to understand and MEMORIZE. I also make some acronyms and short sentences to help you easily remember the credit names.

It should take you about 1 or 2 weeks of full-time study to pass each of the LEED exams. I have met people who have spent 40 hours to study and passed the exams.

You can find sample texts and other information on the LEED Exam Guides series in customer discussion sections under each of my book's listing on Amazon, bn.com and books.google.com.

What others are saying about *LEED GA Exam Guide* (Book 2, LEED Exam Guide series):

"Finally! A comprehensive study tool for LEED GA Prep!

"I took the 1-day Green LEED GA course and walked away with a power point binder printed in very small print—which was missing MUCH of the required information (although I didn't know it at the time). I studied my little heart out and took the test, only to fail it by 1 point. Turns out I did NOT study all the material I needed to in order to pass the test. I found this book, read it, marked it up, retook the test, and passed it with a 95%. Look, we all know the LEED GA exam is new and the resources for study are VERY limited. This one's the VERY best out there right now. I highly recommend it."
—ConsultantVA

"Complete overview for the LEED GA exam

"I studied this book for about 3 days and passed the exam … if you are truly interested in learning about the LEED system and green building design, this is a great place to start."
—K.A. Evans

"A Wonderful Guide for the LEED GA Exam

"After deciding to take the LEED Green Associate exam, I started to look for the best possible study materials and resources. From what I thought would be a relatively easy task, it turned into a tedious endeavor. I realized that there are vast amounts of third-party guides and handbooks. Since the official sites offer little to no help, it became clear to me that my best chance to succeed and pass this exam would be to find the most comprehensive study guide that would not only teach me the topics, but would also give me a great background and understanding of what LEED actually is. Once I stumbled upon Mr. Chen's book, all my needs were answered. This is a great study guide that will give the reader the most complete view of the LEED exam and all that it entails.

"The book is written in an easy-to-understand language and brings up great examples, tying the material to the real world. The information is presented in a coherent and logical way, which optimizes the learning process and does not go into details that will not be needed for the LEED Green Associate Exam, as many other guides do. This book stays dead on topic and keeps the reader interested in the material.

"I highly recommend this book to anyone that is considering the LEED Green Associate Exam. I learned a great deal from this guide, and I am feeling very confident about my chances for passing my upcoming exam."
—Pavel Geystrin

"Easy to read, easy to understand

"I have read through the book once and found it to be the perfect study guide for me. The author does a great job of helping you get into the right frame of mind for the content of the exam. I had started by studying the Green Building Design and Construction reference guide for LEED projects produced by the USGBC. That was the wrong approach, simply too much information with very little retention. At 636 pages in textbook format, it would have been a daunting task to get through it. Gang Chen breaks down the points, helping to minimize the amount of information but maximizing the content I was able to absorb. I plan on going through the book a few more times, and I now believe I have the right information to pass the LEED Green Associate Exam."
—Brian Hochstein

"All in one—LEED GA prep material

"Since the LEED Green Associate exam is a newer addition by USGBC, there is not much information regarding study material for this exam. When I started looking around for material, I got really confused about what material I should buy. This LEED GA guide by Gang Chen is an answer to all my worries! It is a very precise book with lots of information, like how to approach the exam, what to study and what to skip, links to online material, and tips and tricks for passing the exam. It is like the 'one stop shop' for the LEED Green Associate Exam. I think this book can also be a good reference guide for green building professionals. A must-have!"
—SwatiD

"An ESSENTIAL LEED GA Exam Reference Guide

"This book is an invaluable tool in preparation for the LEED Green Associate (GA) Exam. As a practicing professional in the consulting realm, I found this book to be all-inclusive of the preparatory material needed for sitting the exam. The information provides clarity to the fundamental and advanced concepts of what LEED aims to achieve. A tremendous benefit is the connectivity of the concepts with real-world applications.

"The author, Gang Chen, provides a vast amount of knowledge in a very clear, concise, and logical media. For those that have not picked up a textbook in a while, it is very manageable to extract the needed information from this book. If you are taking the exam, do yourself a favor and purchase a copy of this great guide. Applicable fields: Civil Engineering, Architectural Design, MEP, and General Land Development."
—Edwin L. Tamang

Note: Other books in the **LEED Exam Guides series** are in the process of being produced. At least **One book will eventually be produced for each of the LEED exams.** The series include:

LEED GA EXAM GUIDE: *A Must-Have for the LEED Green Associate Exam: Comprehensive Study Materials, Sample Questions, Mock Exam, Green Building LEED Certification, and Sustainability* (3rd Large Format Edition), LEED Exam Guide series, ArchiteG.com (Published January 3, 2011)

LEED GA MOCK EXAMS: *Questions, Answers, and Explanations: A Must-Have for the LEED Green Associate Exam, Green Building LEED Certification, and Sustainability*, LEED Exam Guide series, ArchiteG.com (Published August 6, 2010)

LEED BD&C EXAM GUIDE: *A Must-Have for the LEED AP BD+C Exam: Comprehensive Study Materials, Sample Questions, Mock Exam, Green Building Design and Construction, LEED Certification, and Sustainability* (2nd Edition), LEED Exam Guide series, ArchiteG.com (Published December 26, 2011)

LEED BD&C MOCK EXAMS: *Questions, Answers, and Explanations: A Must-Have for the LEED AP BD+C Exam, Green Building LEED Certification, and Sustainability*, LEED Exam Guide series, ArchiteG.com (Published November 26, 2011)

LEED AP Exam Guide: *Study Materials, Sample Questions, Mock Exam, Building LEED Certification (LEED NC v2.2), and Going Green*, LEED Exam Guides series, LEEDSeries.com (Published on 9/23/2008).

LEED ID&C EXAM GUIDE: *A Must-Have for the LEED AP ID+C Exam: Comprehensive Study Materials, Sample Questions, Mock Exam, Green Interior Design and Construction, LEED Certification, and Sustainability*, LEED Exam Guide series, ArchiteG.com (Published March 8, 2010)

LEED O&M MOCK EXAMS: *Questions, Answers, and Explanations: A Must-Have for the LEED O&M Exam, Green Building LEED Certification, and Sustainability*, LEED Exam Guide series, ArchiteG.com (Published September 28, 2010)

LEED O&M EXAM GUIDE: *A Must-Have for the LEED AP O+M Exam: Comprehensive Study Materials, Sample Questions, Mock Exam, Green Building Operations and Maintenance, LEED Certification, and Sustainability (LEED v3.0)*, LEED Exam Guide series, ArchiteG.com

LEED HOMES EXAM GUIDE: *A Must-Have for the LEED AP Homes Exam: Comprehensive Study Materials, Sample Questions, Mock Exam, Green Building LEED Certification, and Sustainability*, LEED Exam Guide series, ArchiteG.com

LEED ND EXAM GUIDE: *A Must-Have for the LEED AP Neighborhood Development Exam: Comprehensive Study Materials, Sample Questions, Mock Exam, Green Building LEED Certification, and Sustainability*, LEED Exam Guide series, ArchiteG.com

How to order these books:
You can order the books listed above at:
http://www.GreenExamEducation.com

OR
http://www.ArchiteG.com

Building Construction

Project Management, Construction Administration, Drawings, Specs, Detailing Tips, Schedules, Checklists, and Secrets Others Don't Tell You (Architectural Practice Simplified, 2nd edition)

Learn the Tips, Become One of Those Who Know Building Construction and Architectural Practice, and Thrive!

For architectural practice and building design and construction industry, there are two kinds of people: those who know, and those who don't. The tips of building design and construction and project management have been undercover—until now.

Most of the existing books on building construction and architectural practice are too expensive, too complicated, and too long to be practical and helpful. This book simplifies the process to make it easier to understand and uncovers the tips of building design and construction and project management. It sets up a solid foundation and fundamental framework for this field. It covers every aspect of building construction and architectural practice in plain and concise language and introduces it to all people. Through practical case studies, it demonstrates the efficient and proper ways to handle various issues and problems in architectural practice and building design and construction industry.

It is for ordinary people and aspiring young architects as well as seasoned professionals in the construction industry. For ordinary people, it uncovers the tips of building construction; for aspiring architects, it works as a construction industry survival guide and a guidebook to shorten the process in mastering architectural practice and climbing up the professional ladder; for seasoned architects, it has many checklists to refresh their memory. It is an indispensable reference book for ordinary people, architectural students, interns, drafters, designers, seasoned architects, engineers, construction administrators, superintendents, construction managers, contractors, and developers.

You will learn:
1. How to develop your business and work with your client.
2. The entire process of building design and construction, including programming, entitlement, schematic design, design development, construction documents, bidding, and construction administration.
3. How to coordinate with governing agencies, including a county's health department and a city's planning, building, fire, public works departments, etc.
4. How to coordinate with your consultants, including soils, civil, structural, electrical, mechanical, plumbing engineers, landscape architects, etc.
5. How to create and use your own checklists to do quality control of your construction documents.
6. How to use various logs (i.e., RFI log, submittal log, field visit log, etc.) and lists (contact list, document control list, distribution list, etc.) to organize and simplify your work.
7. How to respond to RFI, issue CCDs, review change orders, submittals, etc.
8. How to make your architectural practice a profitable and successful business.

Planting Design Illustrated

A Must-Have for Landscape Architecture: A Holistic Garden Design Guide with Architectural and Horticultural Insight, and Ideas from Famous Gardens in Major Civilizations

One of the most significant books on landscaping!

This is one of the most comprehensive books on planting design. It fills in the blanks of the field and introduces poetry, painting, and symbolism into planting design. It covers in detail the two major systems of planting design: formal planting design and naturalistic planting design. It has numerous line drawings and photos to illustrate the planting design concepts and principles. Through in-depth discussions of historical precedents and practical case studies, it uncovers the fundamental design principles and concepts, as well as the underpinning philosophy for planting design. It is an indispensable reference book for landscape architecture students, designers, architects, urban planners, and ordinary garden lovers.

What Others Are Saying About *Planting Design Illustrated* ...

"I found this book to be absolutely fascinating. You will need to concentrate while reading it, but the effort will be well worth your time."
—**Bobbie Schwartz, former president of APLD (Association of Professional Landscape Designers) and author of *The Design Puzzle: Putting the Pieces Together*.**

"This is a book that you have to read, and it is more than well worth your time. Gang Chen takes you well beyond what you will learn in other books about basic principles like color, texture, and mass."
—**Jane Berger, editor & publisher of gardendesignonline**

"As a longtime consumer of gardening books, I am impressed with Gang Chen's inclusion of new information on planting design theory for Chinese and Japanese gardens. Many gardening books discuss the beauty of Japanese gardens, and a few discuss the unique charms of Chinese gardens, but this one explains how Japanese and Chinese history, as well as geography and artistic traditions, bear on the development of each country's style. The material on traditional Western garden planting is thorough and inspiring, too. *Planting Design Illustrated* definitely rewards repeated reading and study. Any garden designer will read it with profit."
—**Jan Whitner, editor of the *Washington Park Arboretum Bulletin***

"Enhanced with an annotated bibliography and informative appendices, *Planting Design Illustrated* offers an especially "reader friendly" and practical guide that makes it a very strongly recommended addition to personal, professional, academic, and community library gardening & landscaping reference collection and supplemental reading list."
—**Midwest Book Review**

"Where to start? *Planting Design Illustrated* is, above all, fascinating and refreshing! Not something the lay reader encounters every day, the book presents an unlikely topic in an easily digestible, easy-to-follow way. It is superbly organized with a comprehensive table of contents, bibliography, and appendices. The writing, though expertly informative, maintains its accessibility throughout and is a joy to read. The detailed and beautiful illustrations expanding on the concepts presented were my favorite portion. One of the finest books I've encountered in this contest in the past 5 years."
—**Writer's Digest 16th Annual International Self-Published Book Awards Judge's Commentary**

"The work in my view has incredible application to planting design generally and a system approach to what is a very difficult subject to teach, at least in my experience. Also featured is a very beautiful philosophy of garden design principles bordering poetry. It's my strong conviction that this work needs to see the light of day by being published for the use of professionals, students & garden enthusiasts."
—**Donald C. Brinkerhoff, FASLA, chairman and CEO of Lifescapes International, Inc.**

Index

Made in the USA
Lexington, KY
05 January 2015